Root Cause Analysis Handbook

A Guide to Effective Incident Investigation

Risk and Reliability Division
ABS Group Inc.

QUALITY
SAFETY
ABS GROUP
ENVIRONMENT

Government Institutes
Rockville, Maryland

Library of Congress Cataloging-in-Publication Data

Root cause analysis handbook : a guide to effective incident investigation / ABS Group Inc., Risk and Reliability Division
p. cm.
Includes bibliographical references and index.
ISBN 0-86587-658-4 (paper)
1. Quality control--Data processing. 2. Industrial accidents--Investigation. 3. Critical incident technique. I. ABS Group Inc., Risk and Reliability Division.
TS156.R665 1999
658.5'62--dc21 99-22525
CIP

Printed in the United States of America

Thank You for Choosing the *Root Cause Analysis Handbook* as Your Root Cause Analysis Resource.

If You Are Looking for More Help...

ABS Group Inc. personnel have worked on all types of root cause analyses and incident investigations. These range from identifying human errors or component failures that contribute to simple system failures to discovering the origins of catastrophic events by piecing together a complex chain of events through rigorous application of root cause analysis to analyzing chronic failures at your facility. Our techniques have been applied to industrial accidents, production bottlenecks, reliability problems, quality concerns, and financial issues.

ABS Group Inc. Investigation Assistance

If you need help investigating an accident or problems related to reliability, quality, or finances, ABS Group Inc. can assist with your analysis of the problem. We have worked with organizations in a variety of industries to investigate chronic problems associated with reliability, quality, and financial performance, and industrial accidents and fatalities.

ABS Group Inc. Training Services

Based on our experiences, we have trained thousands of individuals using the proven techniques outlined in this handbook. And because these courses emphasize a workshop approach to learning, students gain valuable experience by practicing what they learn on realistic industry examples. We can even customize these workshops to make them specific to your company or facility. The courses can range from 1 to 5 days in duration. Following are summaries of some of our standard public courses.

Incident Investigation (Course 106) — The focus of this course is on how to gather data, analyze data for causal factors, fill gaps in data, determine root causes, and write effective recommendations. You will also learn how to structure an effective incident investigation program, which includes defining, classifying, and trending data on near misses and other incidents that need to be reported in your program.

Root Cause Analysis (Course 123) — In this course you will learn and apply several proven, systematic methods (such as the 5 Whys and Change Analysis) to uncover the root causes behind system performance problems. You will also participate in several workshops, including one on the use of ABS Group Inc.'s Root Cause Map™ and a complete, simulated root cause analysis of a realistic reliability problem.

Preventing Human Error (Course 124) — In this course you will learn how to examine human errors to identify the conditions and *error-likely situations* that contributed to the mistake. From this starting point, you will learn to recognize the true causes of most human errors — weaknesses in the management systems we use to (1) design equipment and processes, (2) develop and use procedures and policies, and (3) select, train, supervise, and communicate with workers.

Component Failures (Course 208) — This course introduces the many fundamental mechanisms of component failures. You will learn the basic steps that experts follow when analyzing component failures and how typical failure mechanisms apply to the most common components in manufacturing processes.

Contact Us for Information and Assistance

If you would like a catalog of our complete line of courses or more information about how our analysis services could be used at your facility, contact ABS Group Inc.

- by phone at **1-865-966-5232**,
- by fax at **1-865-966-5287**,
- by e-mail at **info@abs-jbfa.com**, or
- through the world wide web at **http://www.abs-jbfa.com**.

You may also copy, complete, and fax the form below (or mail it to us at **ABS Group Inc. Risk & Reliability Division, 1000 Technology Drive, Knoxville, Tennessee, 37932, USA**).

Information Request Form

__ Please send me your latest training catalog.

__ Please send me more information about your incident investigation and root cause analysis services.

Name: ______________________________

Title: ______________________________

Company: ______________________________

Address: ______________________________

City: ______________________________

State: ______________________________

Zip: ______________________________

Phone: ______________________________

Fax: ______________________________

E-mail: ______________________________

Table of Contents

Organization of the Root Cause Analysis Handbook vii

Acronyms and Nomenclature ix

Limitations of Liability xi

Acknowledgements xiii

Chapter 1 Introduction to Root Cause Analysis 1

Chapter 2 Collecting and Preserving Data for Analysis 9

Chapter 3 Data Analysis Using Causal Factor Charting 11

Chapter 4 Root Cause Identification 25

Chapter 5 Recommendation Generation and Implementation 33

Chapter 6 References 37

Appendix A Root Cause Map™ Indexes Node Descriptions ..... A-i

Appendix B Root Cause Map B-1

Organization of the Root Cause Analysis Handbook

The focus of this handbook is on the application of the Root Cause Map™ to the root cause analysis process. The Root Cause Map is used in one of the later steps of the root cause analysis process to identify the underlying management systems that caused the event to occur or made the consequences of the event more severe. The first five chapters of this handbook are an overview of the root cause analysis process. These provide the context for use of the Root Cause Map. Chapter 6 provides references.

Chapter 1, "Introduction to Root Cause Analysis," presents a basic overview of the SOURCE™ (Seeking Out the Underlying Root Causes of Events) root cause analysis process. Chapter 2, "Collecting and Preserving Data for Analysis," outlines the types of data and data sources that are available. Chapters 3, 4, and 5 describe the three major steps in the root cause analysis process. Chapter 3, "Data Analysis Using Causal Factor Charting," provides a step-by-step description of causal factor charting techniques. Chapter 4, "Root Cause Identification," explains the organization and use of the Root Cause Map. Chapter 5, "Recommendation Generation and Implementation," provides guidance on developing and implementing corrective actions. The references section, Chapter 6, provides additional information for those interested in learning more about specific items contained in the handbook.

Appendix A, "Root Cause Map Node Descriptions," describes each segment of the Root Cause Map and presents detailed descriptions of the individual nodes on the map. Appendix B is the Root Cause Map itself.

Acronyms and Nomenclature

ABS Group	ABS Group Inc. Risk & Reliability Division
CF	causal factor
DG	diesel generator
EPA	Environmental Protection Agency
FMEA	failure modes and effects analysis
HAZOP	hazard and operability analysis
LTA	less than adequate
OSHA	Occupational Safety and Health Administration
P&ID	piping and instrumentation diagram
PPE	personal protective equipment
PSSR	pre-startup safety review
SOURCE™	Seeking Out the Underlying Root Causes of Events
SPACs	standards, policies, or administrative controls

Limitations of Liability

ABS Group Inc. Risk & Reliability Division
1000 Technology Drive
Knoxville, TN 37932-3353

Acknowledgements

ABS Group Inc. Risk & Reliability Division thanks the many personnel who contributed to the development and ongoing revision of this handbook, particularly its primary author, Lee N. Vanden Heuvel. We also thank Leslie K. Adair, William G. Bridges, Rebekah Ellis, Donald K. Lorenzo, Randal L. Montgomery, Steven G. Schoolcraft, and Tom R. Williams for reviewing this handbook. And we thank Jill M. Johnson, Nicole M. Lepoutre-Baldocchi, Paul M. Olsen, Robin M. Ragland, and Maleena L. Wright for their skill and craftsmanship in preparing this handbook. We are also grateful for the support and assistance of the rest of the staff at ABS Group Inc. Risk & Reliability Division.

CHAPTER 1

INTRODUCTION TO ROOT CAUSE ANALYSIS

Objectives and Scope

The root cause analysis system presented in this handbook is designed for use in investigating and categorizing the root causes of events with safety, health, environmental, quality, reliability, and production impacts, although the examples used in this handbook are predominantly those having safety and health impacts. The term "event" is used to generically identify events that have these types of consequences. The SOURCE™ (Seeking Out the Underlying Root Causes of Events) methodology is based on one developed for the Department of Energy by the Westinghouse Savannah River Corporation in 1991.

Root cause analysis is simply a tool designed to help investigators (1) describe WHAT happened during a particular occurrence, (2) determine HOW it happened, and (3) understand WHY it happened. Only when investigators are able to determine WHY an event or failure occurred will they be able to specify workable corrective measures.

Most event analysis systems allow investigators to answer questions about what happened during an event and about how the event occurred, but often they are not encouraged to determine why the event occurred. Imagine an occurrence during which an operator is instructed to close Valve A; instead, the operator closes Valve B. The typical investigation would probably result in the conclusion that "operator error" was the cause of the occurrence. This is an accurate description of what happened and how it happened. An operator committed an error by manipulating the wrong valve. If the analysts stop at this level of analysis, however, they have not probed deeply enough to understand the reasons for the mistake. Generally, mistakes do not "just happen." They can be traced to some well-defined causes. In the case of the valving error, we might ask: Was the procedure confusing? Were the valves clearly labeled? Was the operator who made the mistake familiar with this particular task? These and other questions should be asked to determine why the error took place.

When the analysis stops at the point of answering WHAT and HOW, the recommendations for preventing recurrence of the event may be deficient. In the case of the operator who turned the wrong valve, we are likely to see recommendations like "Retrain the operator on the procedure," "Remind all operators to be alert when manipulating valves," or "Emphasize to all personnel that careful attention to the job should be maintained at all times." Such recommendations do little to prevent future occurrences. Investigations that probe more deeply into WHY the operator error occurred are able to provide more specific, concrete, and effective recommendations. In the case of the valving error, examples might include, "Revise the procedure so that references to valves match the valve labels found in the field" or "Require operator trainees to have a training procedure in hand when manipulating valves."

The SOURCE root cause analysis system provides a structured approach for the investigators trying to discover the WHYs surrounding a particular occurrence. Identifying these root causes is the key to preventing similar occurrences in the future. An added benefit of an effective root cause analysis is that, over time, the root causes identified across the population of occurrences can be used to target major opportunities for improvement. For example, if a significant number of analyses point to procedure inadequacies as root causes, then resources can be focused on procedure improvement programs. Trending of root causes allows tracking of occurrence causes, development of systematic improvements, and assessment of the impact of corrective programs.

This handbook does not address programmatic issues, such as how to classify events, the definitions of accidents and near misses, how and whom to notify following an event, how to determine team composition, follow-through on results and recommendations, trending, etc. Figure 1-1 illustrates the overall event analysis process; this handbook covers the topics indicated in the figure.

Definition of Root Cause

Although there is substantial debate concerning the definition of a root cause, the SOURCE methodology uses the following definition: *Root causes are the most basic causes that can reasonably be identified, which management has control to fix and for which effective recommendations for preventing recurrence can be generated.* This definition contains the following four key elements:

Root causes are basic causes.

The investigator's goal should be to identify basic causes. The more specific the investigator can be about the reasons why an event occurred, the easier it will be to arrive at recommendations that will prevent recurrence of the events leading up to the occurrence.

Root causes are those causes that can reasonably be identified.

Occurrence investigations must be completed within a reasonable time frame. It is not practical to keep valuable manpower indefinitely occupied searching for the root causes of occurrences. Root cause analysis helps analysts get the most out of the time they have allotted for the investigation.

Root causes are those causes over which management has control.

Analysts should avoid using general cause classifications such as "operator error." Such causes are not specific enough to allow those in charge to rectify the situation. Management needs to know exactly why a failure occurred before action can be taken to prevent recurrence.

Root causes are those causes for which effective recommendations can be generated.

Recommendations should directly address the root causes identified during the investigation. If the analysts arrive at vague recommendations such as "Remind operator to be alert at all times," then they probably have not found a basic enough cause and need to expend more effort in the analysis process.

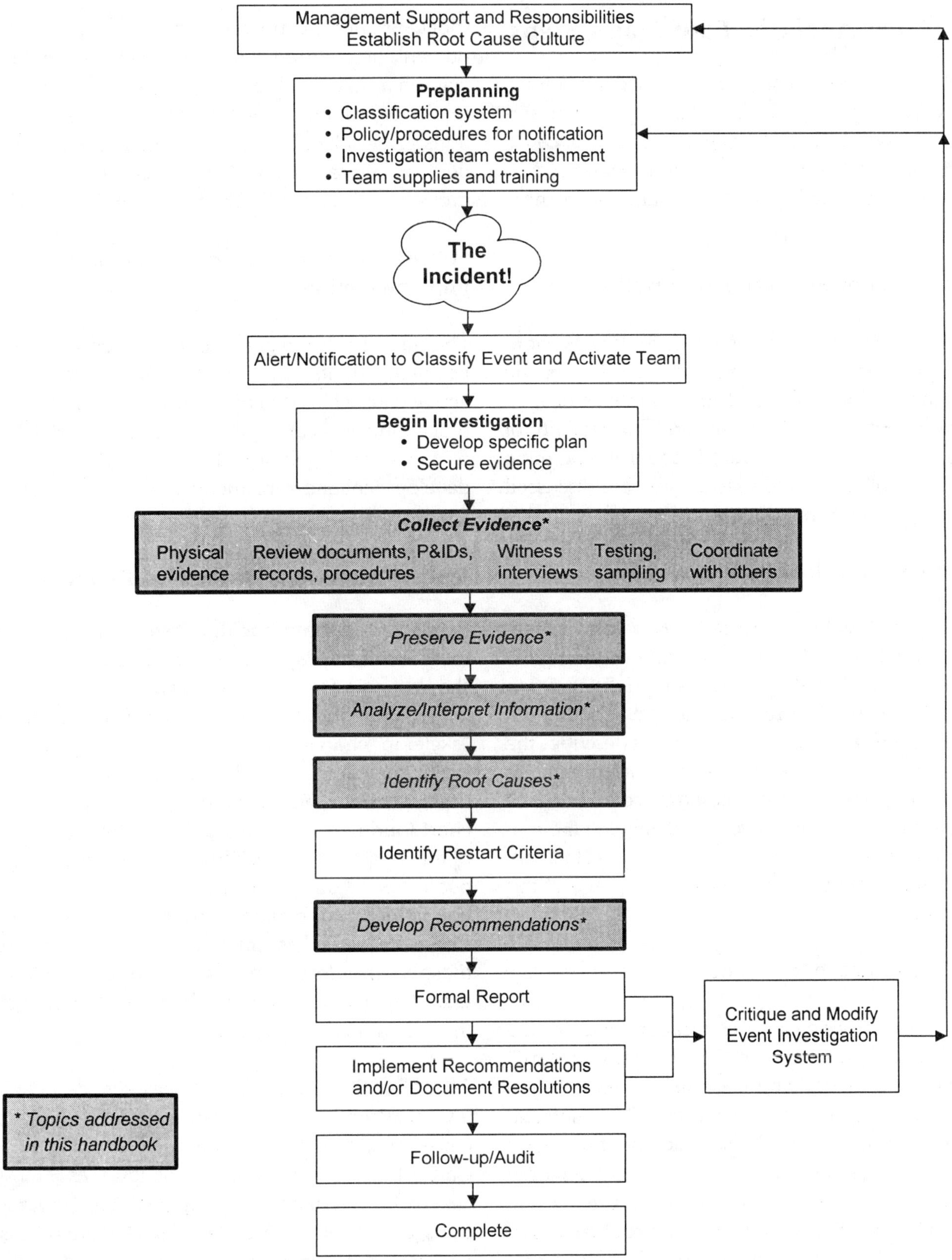

Figure 1-1 Overall Event Investigation Process

Root Cause Analysis: Four Major Steps

The SOURCE root cause analysis process is a four-step process involving: (1) data collection and preservation, (2) causal factor (CF) charting, (3) root cause identification, and (4) recommendation generation and implementation. See Figure 1-2 for a summary of these steps.

Step 1: Data Collection and Preservation

The first step in the analysis is to gather data. Without complete information and an understanding of the event, the causal factors and root causes associated with the event cannot be identified. The majority of time spent analyzing an event is spent in gathering data. Data collection and preservation are discussed in Chapter 2 of this handbook.

Step 2: Causal Factor Charting

CF charting provides a way for investigators to organize and analyze the information gathered during the investigation and to identify gaps and deficiencies in knowledge as the investigation progresses. The CF chart is simply a sequence diagram that describes the events leading up to and following an occurrence, as well as the conditions surrounding these events. The final step in CF charting involves identifying the major contributors to the occurrence (i.e., causal factors). CF charting is discussed in detail in Chapter 3 of this handbook.

Step 3: Root Cause Identification

The next step, root cause identification, involves the use of a decision diagram called the Root Cause Map™ to identify the underlying reason(s) for each causal factor identified during CF charting. The identification of root causes helps the investigator of a specific event determine the reasons why the event occurred so that the problems surrounding the occurrence can be fixed. In addition, trending of the root causes of occurrences identified over a period of time can provide valuable insight concerning specific areas for improvement. This is an added benefit of the SOURCE root cause analysis process. Not only can it be used to prevent the recurrence of specific events, but lessons learned from individual occurrences can be combined to identify major areas of weakness. This allows actions to be taken before a seemingly unrelated accident or failure occurs. The root cause identification process is discussed in Chapter 4 of this handbook.

Step 4: Recommendation Generation and Implementation

The next step is the generation of recommendations. Following identification of the root cause(s) for a particular causal factor, achievable recommendations for preventing its recurrence must be generated. Chapter 5 of this handbook provides some guidelines for the development and implementation of adequate corrective actions.

Root Cause Analysis: The Process

Figure 1-3 illustrates the major steps in the SOURCE root cause analysis process. Preparation of the CF chart should begin as soon as investigators start to collect information about the occurrence. They begin with a "skeleton" chart that is modified as more relevant facts are uncovered. Data collection continues until the investigators are satisfied with the thoroughness of the chart (and hence are satisfied with the thoroughness of the investigation). When the entire occurrence has been charted out, the investigators are in a good position to identify the major contributors to the incident. These are labeled as causal factors. *Causal factors are those contributors (human errors and component failures) that, if eliminated, would have either prevented the occurrence or reduced its severity.*

After all of the causal factors have been identified, the investigators begin root cause identification. Each causal factor is analyzed, one at a time, using the Root Cause Map. The map structures the reasoning process of the investigators by helping them answer questions about why particular causal factors exist or occurred. After each causal factor is analyzed, the investigators attempt to arrive at recommendations that will prevent its recurrence. This process continues un-

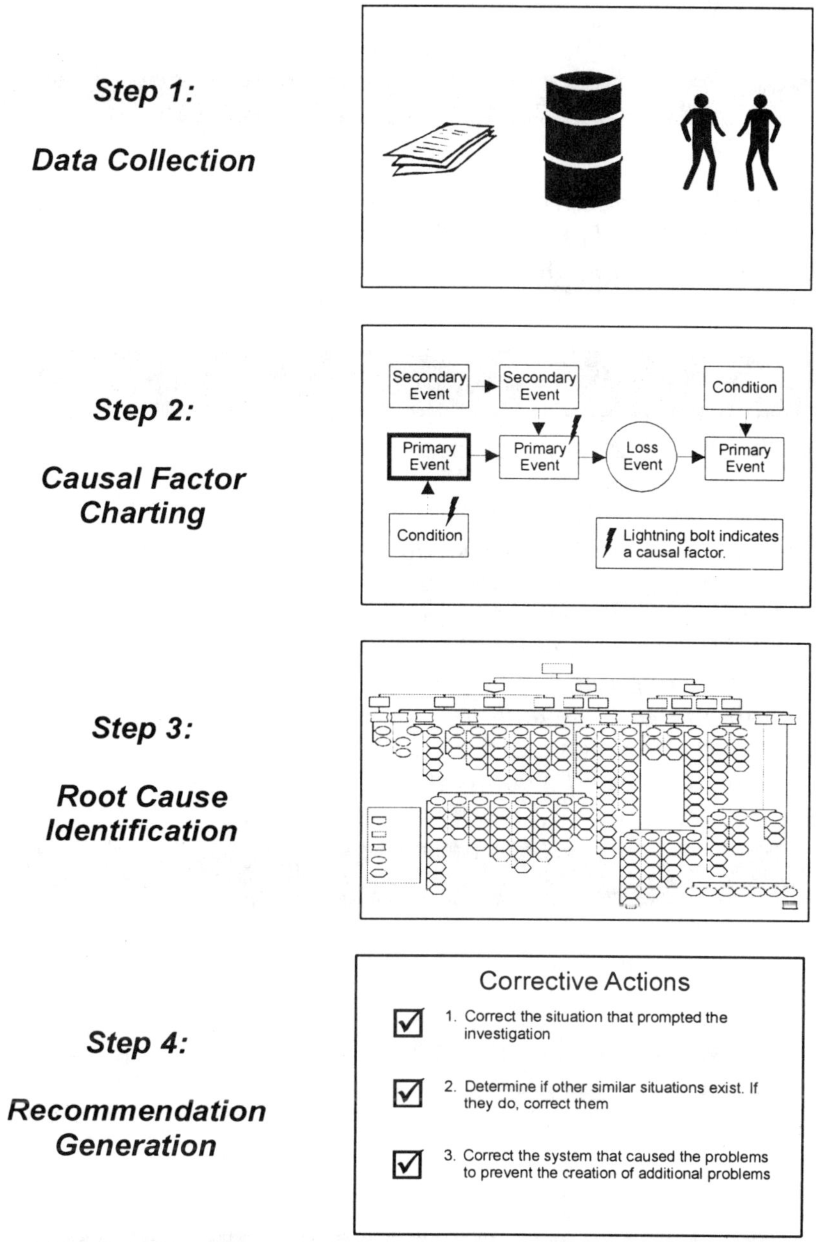

Figure 1-2 The Four Steps of Root Cause Analysis

til root causes have been identified for each causal factor.

In many traditional analyses, the most visible causal factor is given all of the attention. Often, the investigators are tempted to "jump to conclusions" about how to solve the problem. Rarely are events caused by one causal factor. They are usually the result of a combination of contributors. When only one predominant causal factor is addressed, the list of recommendations will likely not be complete. Consequently, the occurrence may repeat itself. To help prevent the analyst from omitting important recommendations, the SOURCE root cause analysis process requires that all causal factors be determined from analysis of the relevant events data and that each causal factor be addressed separately. Root causes are identified for each causal factor, and recommendations are developed to address each root cause. When recommendations are generated in this manner, one at a time, the probability of missing important details decreases.

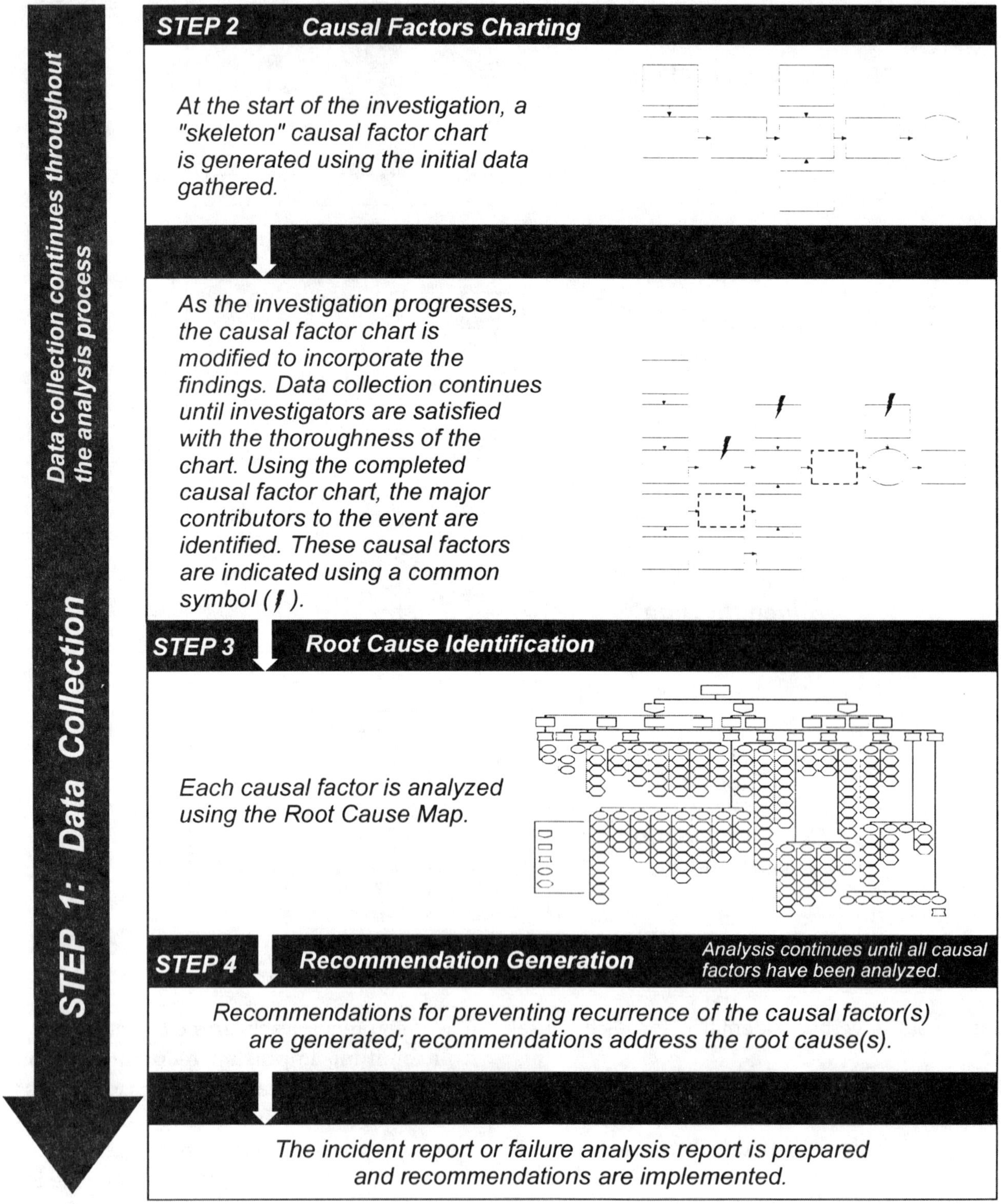

Figure 1-3 The Analysis Process

Root Cause Analysis: Presentation of Results

The SOURCE methodology uses root cause summary tables, such as the one presented in Figure 1-4, to organize the information compiled during CF charting, root cause identification, and recommendation generation. A summary table is prepared for each causal factor identified during CF charting. The table is divided into three columns with each column representing a major aspect of the root cause analysis process (i.e., identification of a causal factor, root cause identification, and recommendation generation). In the first column, a general description of the causal factor is presented. This column provides sufficient detail for the reader of an occurrence report to be able to understand, in a general sense, the scenario surrounding the causal factor. The second column shows the path or paths through the Root Cause Map that were used to categorize the causal factor. The third column presents recommendations to address each of the root causes identified for the causal factor. Use of this three-column format aids the investigator in addressing each causal factor individually and is effective in ensuring that all important items are sufficiently covered.

The end result of a root cause analysis investigation is generally an investigation report. Reporting formats vary according to the particular reporting system requiring the investigation. The format of the report is generally well defined by the administrative documents governing the particular reporting system; however, a thorough root cause analysis will greatly simplify the preparation of any type of investigative report. The completed CF chart provides an excellent basis for the occurrence description required by most reporting systems. Root cause identification should leave the investigators feeling confident that they have discovered the reasons why the event occurred. In addition, a quick check for obvious blank spaces in the root cause summary tables should help ensure that the investigators have generated recommendations for each root cause.

If the investigators have completed a table for each of the causal factors identified, then the results of the root cause analysis are completely documented. Although the internal requirements of a company for an event report format may not be flexible enough to allow the complete root cause analysis to be placed in the body of the occurrence report, it is usually appropriate to attach the CF chart and the tables as appendices to the final document.

The root cause analyst is often not responsible for the implementation of recommendations generated by the analysis. However, if the recommendations are not implemented, the effort expended in performing the analysis is wasted. In addition, the events that triggered the analysis should be expected to recur. Because the recommendations are not implemented, the situation has not been changed and it is inevitable that the event will occur again.

Summary

The goal of root cause analysis is not only to understand the what and how of an event, but also why it happened. The analysis of an event begins with the gathering of data. As the data is gathered, it is organized and analyzed using causal factor charting. The goal is to identify the causal factors for the event. Causal factors are those contributors (human errors and component failures) that, if eliminated, would have either prevented the occurrence or reduced its severity. Once the event is understood by using causal factor charting and other analysis techniques, root causes are identified for each causal factor. Root causes are the most basic causes that can reasonably be identified, which management has control to fix and for which effective recommendations for preventing recurrence can be generated. Finally, recommendations are developed and implemented to prevent the causal factors from occurring again.

Step 1

Step 2

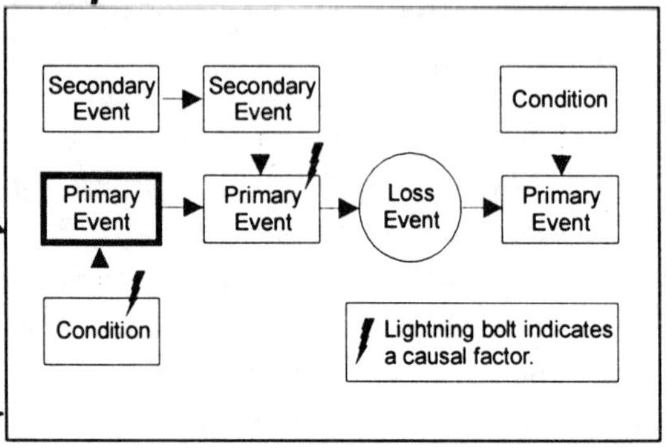

Title of Event Investigation Report		
Causal Factor #1	**Paths Through Root Cause Map**	**Recommendations**
In the first space in column one, repeat the causal factor as it appears in the causal factor chart. **BACKGROUND:** In this space, provide a brief background that explains the causal factor listed, including underlying reasons why the causal factor occurred or was allowed to exist. This space could also be used to elaborate on the causal factor, if the causal factor statement is terse.	• Primary difficulty source • Problem category • Root cause category • Near root cause • Root cause In this space, list the paths through the map to a root cause.	In this column, list (or reference) any recommendations (or decision not to make a recommendation) to address each root cause.
	• Primary difficulty source • Problem category • Root cause category • Near root cause • Root cause List the paths for any other root cause(s).	List recommendations for the other root causes associated with this causal factor.

Step 3

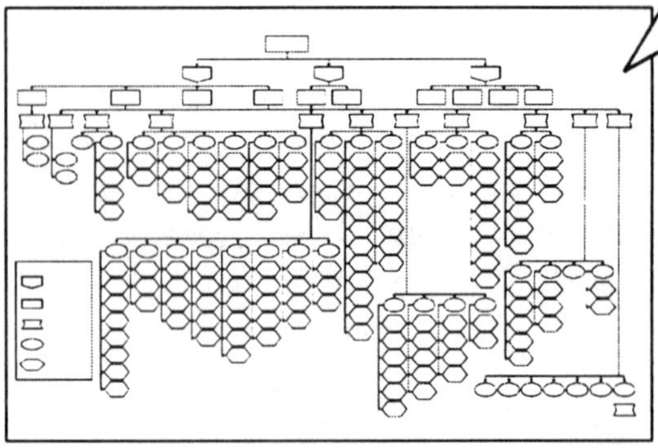

Step 4

Corrective Actions

☑ 1. Correct the situation that prompted the investigation

☑ 2. Determine if other similar situations exist. If they do, correct them

☑ 3. Correct the system that caused the problems to prevent the creation of additional problems

Figure 1-4 Root Cause Summary Table

CHAPTER 2

COLLECTING AND PRESERVING DATA FOR ANALYSIS

Factual evidence derived from data-gathering activities serves as the basis for all valid conclusions and recommendations from a root cause analysis. Without effective data gathering, the event (problem) cannot be truly defined and solved. Data gathering is an ongoing process throughout the analysis. It is the first step of the process and continues while causal factors and root causes are identified.

There are a number of different types and sources of data. The four basic types of data are:

People — witnesses, participants, etc.
Physical — parts, chemical samples, etc.
Position — location of people and physical evidence
Paper — both hard copies and software versions

Data gathering must begin as soon as possible after the event occurs to prevent loss or alteration of the data. Data from people are the most fragile; that is, most easily altered or destroyed. Gathering data from people needs to be the first priority. Other forms of data are more stable; however, physical data need to be identified quickly to prevent their inadvertent destruction.

To gather data from people, the analyst must be a skilled interviewer. Figure 2-1 describes the overall interview process. Planning for the interview is essential if the interviewer wants to get the maximum amount of data from the interview. During the interview, the interviewer must ask open-ended questions – questions that require the interviewee to respond with a long, descriptive answer. After these questions are asked and answered, the interviewer can follow up with more detailed questions. At the conclusion of the interview, the analyst will use the data gathered to update the causal factor chart and/or the fault tree (see Chapter 3).

If investigators cannot arrive at the event scene quickly, the witnesses should complete initial witness statements. These will help during the interview process by providing the interviewer with an outline of the information the witnesses can provide.

Follow-up interviews may be needed to answer additional questions that arise during the course of the analysis. These interviews are conducted in the same general manner as the initial interviews, but a more structured, straight-to-the-point interview style is desirable.

The second type of data are physical data. Physical data consist of parts, residues, and chemical samples. To analyze physical data, the data must first be identified and preserved. This is to prevent inadvertent destruction of the data. Test plans should also be used for analysis of physical data. Test plans are developed to ensure complete collection of required data and complete analysis of the evidence, and to prevent inadvertent destruction of evidence by the investigators.

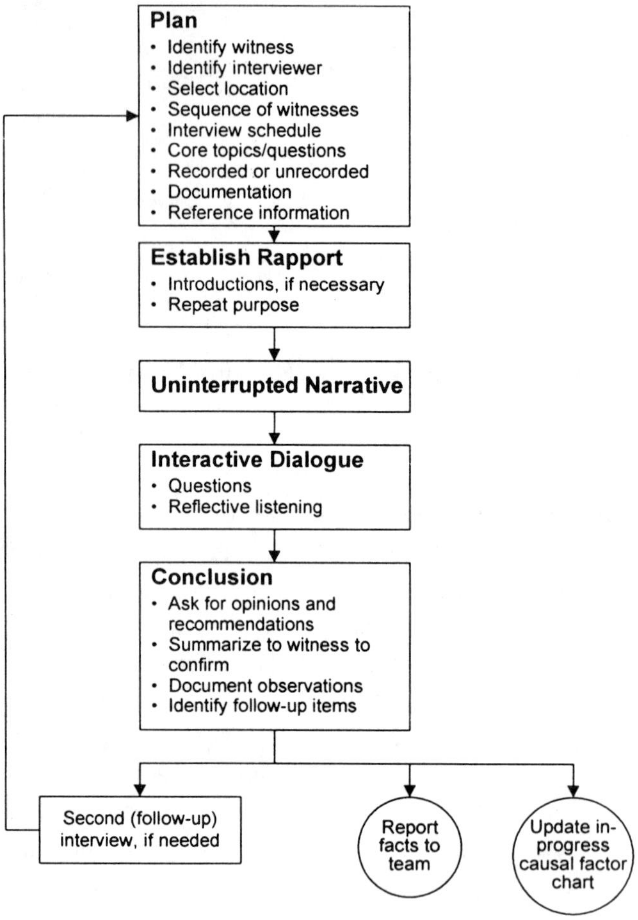

Figure 2-1 Interview Process

The third type of data are position data. Position data consist of:

- Physical relationships among items and people at the scene and environmental factors
- Functional relationships among control devices and safety devices
- Time relationships that will define the sequence of events and help identify cause-effect relationships

Physical position data can be obtained by using photography to record physical position information. Functional relationships among systems and time relationships can be understood by analyzing other sources of data.

The last type of data are paper data. Paper data consist of data on paper and data stored electronically that can printed out on paper. Examples of paper data include documentation records, logs and data-recording results, procedures, memos, correspondence, program manuals, and policy statements. Paper data can often lead to root causes because many management systems are documented on paper.

The causal factor chart and fault tree techniques covered in the next chapter will help guide the data analysis process.

Summary

Factual evidence derived from data-gathering activities serves as the basis for all valid conclusions and recommendations generated by a root cause analysis. Data gathering is an ongoing process throughout the analysis. The four basic types of data are: people, physical, position, and paper. The most fragile type of these is people data. Data must be gathered quickly to prevent their inadvertent destruction.

CHAPTER 3

DATA ANALYSIS USING CAUSAL FACTOR CHARTING

When an investigator or investigation team begins an analysis, the analyst uses a causal factor chart or fault tree to organize and analyze the data. A CF chart is simply a sequence diagram that allows investigators to graphically depict the mishap from beginning to end. The CF charting technique was originally developed by Ludwig Benner and his colleagues at the U.S. National Transportation Safety Board for use as an analytical tool in accident investigations. The tool is designed to help investigators chronologically describe the events leading up to an unwanted occurrence and the conditions surrounding these events.

Fault trees are a structured process for postulating ways the system could fail. The tree is developed only in enough detail to understand how the system failed.

Benner (1975) suggests that an accident involves a sequence of events (i.e., happenings) that occur during the course of good-intentioned work activity but that culminate in unintentional personnel injury or damage to a system. Experience has shown that incidents develop from clearly defined sequences of events that involve performance errors, changes, oversights, and omissions. The event investigator needs to identify and document not only these negative events, but also the relevant conditions and nonhazardous events related to the incident sequence. CF charts and fault trees are excellent vehicles for accomplishing this purpose.

The CF Chart: Definition of Elements

The principles of CF charting are quite basic. Figure 3-1 presents a sample CF chart, and Table 3-1 provides definitions of the CF chart elements. Notice that the sample CF chart is constructed of a number of different building blocks.

The objective of creating building blocks is to transform observations and data acquired from people and things into a common format needed to construct the CF chart. A building block is a fundamental, common, and irreducible unit that describes events/conditions of an event. These building blocks are used to chart the sequence of events and conditions that led to the incident under investigation. This procedure will work with any observation or data about any occurrence.

Events and Conditions

The most basic elements of a CF chart are events and conditions. Events are simply the *actions* or *happenings* that occur during some sequence of activity. Events make up the backbone of the CF chart. Event statements describe specific occurrences (e.g., "4-12 shift operator filled Tank 123" or "Control room operator acknowledged level alarm for Tank 57").

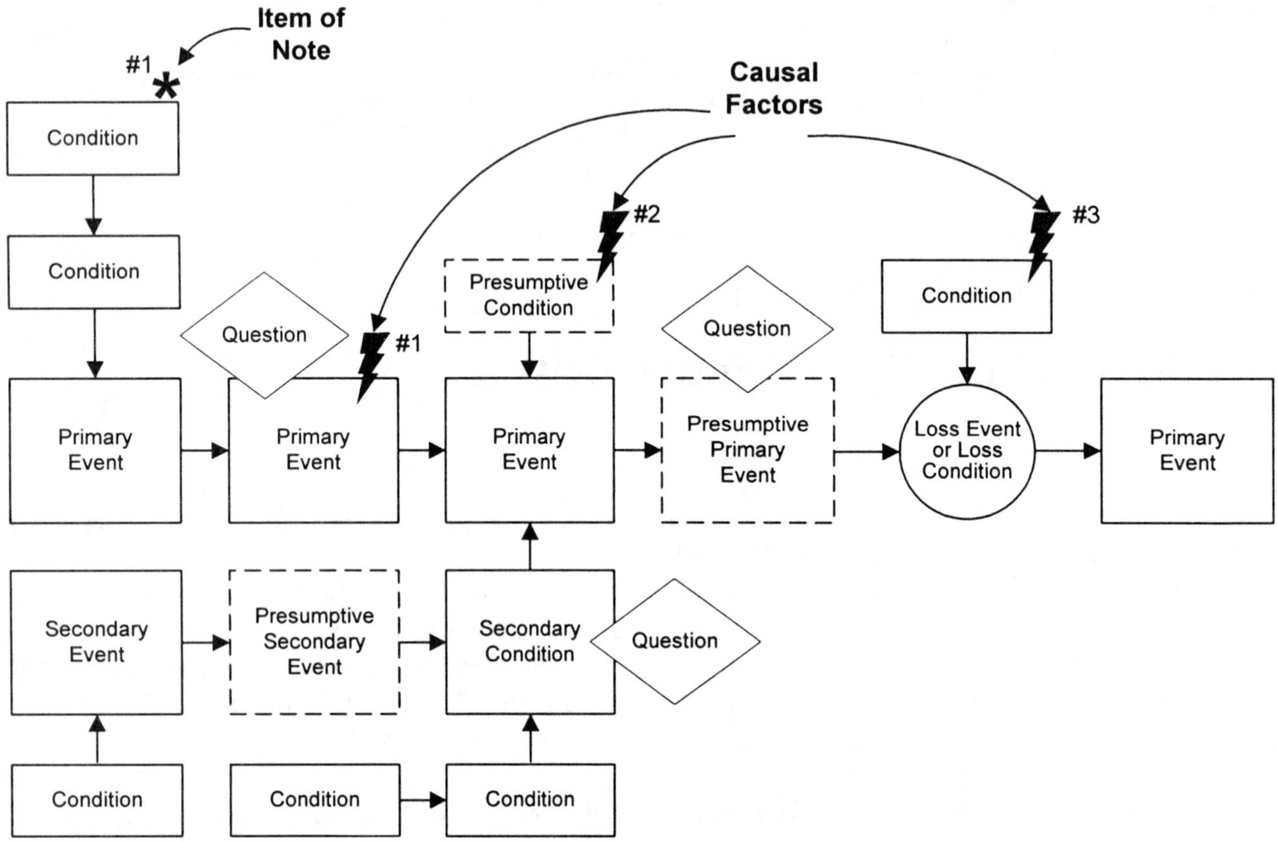

Figure 3-1 Sample Causal Factor Chart

Conditions are not specific activities, but circumstances pertinent to the situation. Conditions usually provide descriptive information (e.g., "Pressure was 1,000 psig") as opposed to stating action (e.g., "Operator placed Valve X into open position"). Conditions typically describe a state of the system, a value of a system, or an environmental parameter. Conditions can also be used to state background information (such as the level of training of an operator) or to summarize the state of the system at some point in the CF chart (such as "The line is now completely open at the decanter").

Loss Events

The event(s) that describes the negative consequences of the event is referred to as the loss event. Another way to think of the loss event is that it probably describes the reason an investigation is required (e.g., personnel injuries, fires, explosions, release of materials, unscheduled shutdowns of equipment, scrapped product). Loss events usually appear on the right-hand side of the causal factor chart.

Questions

Questions will be placed on the chart to identify gaps in the understanding of the analysis to guide further data-gathering activities. For each question, a potential source of data that can resolve the issue is also identified. This allows the investigator to prepare for interviews of people and to prepare test plans for the analysis of physical data.

Facts Versus Suppositions

During the analysis, there may be some gaps in the data that are gathered that simply cannot be filled in. The data that can fill in these gaps may no longer be available, or the cost of obtaining the data may be too

Table 3-1 Definitions of CF Chart Elements

Elements	Definitions
Loss Event	A statement of the negative consequence or "near hit"; the event that necessitates the event investigation
Primary Events	Actions or happenings leading up to the loss event. These can be negative events or expected events
Secondary Events	Actions or happenings that explain why the primary events occurred
Presumptive Events	Actions or happenings, not based upon *valid** factual evidence, that are assumed because they appear logical in the sequence of events
Conditions	Circumstances pertinent to the situation; usually provide descriptive information, such as parametric conditions (can also be used to denote summary statements or conclusions)
Loss Condition	A statement of the hazard condition that could have (but did not) result in the negative consequence; the condition that necessitates the event investigation
Presumptive Conditions	Circumstances, not based upon *valid** factual evidence, that are assumed because they appear logical in the sequence of the events (can also be used to denote summary statements or conclusions)
Causal Factors	Human errors and equipment failures that, if eliminated, would have either prevented the occurrence or reduced its severity
Presumptive Causal Factors	Causal factors, not based upon *valid** factual evidence, that are logically assumed to be major contributors to the event
Items of Note	Significant deficiencies identified during the course of the investigation that were not major contributors to the event, but should be addressed before they have the opportunity to cause problems

* The definition of *valid* is subjective; usually if team members agree on the validity of a data item, we call it a fact. Statements from witnesses or items of data from computer control systems are usually accepted as *valid facts*.

high to justify its gathering. These gaps may be filled in on the chart using suppositions. Suppositions are good guesses about what happened based on all the other facts that have been collected. It is important to differentiate between the facts and the suppositions that are shown on the chart. By doing so, it is clear to the reader what items on the chart are facts and what information on the chart are suppositions.

Some events and conditions, although they may appear to be logical in the sequence of the diagram, cannot be substantiated with valid factual evidence. These are referred to as presumptive events and presumptive conditions (or can simply be referred to as suppositions). On the CF chart, events and conditions of this type are clearly distinguished from elements based on hard facts (such as by using a dashed outline instead of a solid outline).

One of the first decisions the investigator will have to make is whether the event is based on fact or supposition. Uncertainty in the data collected from witnesses and things is common. This uncertainty arises for many reasons, including the mind set of witnesses, the motives of witnesses, deterioration of the data source, or gaps in the data. The investigator can choose to create a building block for data with uncertainty (i.e., a supposition), but a clear distinction should be made between a fact and a supposition. Investigators commonly feel uncomfortable calling any item of data a *fact*, since all such judgments are subjective.

When developing building blocks, a simple rule of thumb is to treat any of the following as *facts*:

- Direct observation by a witness (including memory of time and field readings)
- Data recorded by process instruments
- Strong conclusions reached by those who analyze physical evidence such as parts/equipment or residual chemicals
- Paper evidence (to the extent that the program, procedure, or record relates to the process in which the incident occurs)

Examples of *suppositions* include:

- Second-hand testimony or hearsay
- Conclusions by witnesses or others (except as stated above)

Some statements by witnesses may be classified as either a fact or a supposition, depending on the perspective of the investigator(s). For example, in the statement "It was very bright that day," the conclusion drawn by the witness of the day being very bright may be accepted as a fact if all interested parties are expected to understand the qualitative meaning of "very bright." In most cases such as this, the distinction between fact and supposition is of minor importance and probably not worth debate by investigation team members. However, if this issue affects the recommended corrective actions, or if the investigation results will be used in a court case or in dealing with regulators on potential citations or judgments, then clearly differentiating between facts and supposition is very important.

Causal Factors and Items of Note

The goal of data analysis is to identify the key human errors and equipment failures that led to or allowed the loss event to occur, or increased the size of the loss event. These are called causal factors. In the next step of the process, the root causes of each causal factor will be identified. Often the investigation will uncover deficiencies that did not contribute to the incident, but which must be addressed before they have the opportunity to cause problems. Often times these deficiencies will be relatively easy to fix; however, in other instances, they may represent large programmatic deficiencies. These items should not be identified as causal factors because they did not contribute directly to the loss event; however, it is important to highlight them on the CF chart as items of note and to generate recommendations addressing them.

Causal Factors

After the CF chart has been completed, the investigators are in a good position to identify factors that influenced the course of events. These elements are labeled causal factors. Causal factors, which may be in the form of negative events or hazardous conditions, are those items that are considered major contributors to the occurrence. Elimination of these casual factors would have either prevented the occurrence or reduced its severity.

Irrelevant Data

An investigator collects considerable data during a thorough investigation. Information (data) overload can cloud the real issues and impede the progress of an investigation. With experience, an investigator will learn to recognize the difference between important data and extraneous, irrelevant, or redundant data. A goal of the investigator should be to create **one** building block for each **relevant** data item. Examples of irrelevant data might be the color of a switch in a chemical process. The color of the switch may not be relevent to understanding why the operator selected the wrong switch. However, if the switches are color coded, the color becomes important in understanding the human error associated with operating the switch. Common sense and experience should be used by investigators to ensure that their limited investigation resources and efforts are focused on determining cause-effect relationships directly related to the actual incident or similar incidents. Events charting and deductive reasoning analysis techniques (using the building blocks) will help guide the investigator in the search for relevant data.

The procedure for formulating building blocks from any kind of data requires that the "actor/parameter + verb" format be **rigorously** followed. During an investigation, investigators transform *all relevant* information they gather from people or things into this format, and then document it as building blocks. (Note that *all* can be taken to extremes. Much of the data collected may be irrelevant to the scope of the investigation or redundant, as discussed earlier.) To record investigative information, use self-stick notes (such as Post-It® Notes) of 3" x 3" size. Have at least two different colors on hand.

A very simple building block is illustrated in Figure 3-2. Figure 3-3 shows a comprehensive building block .

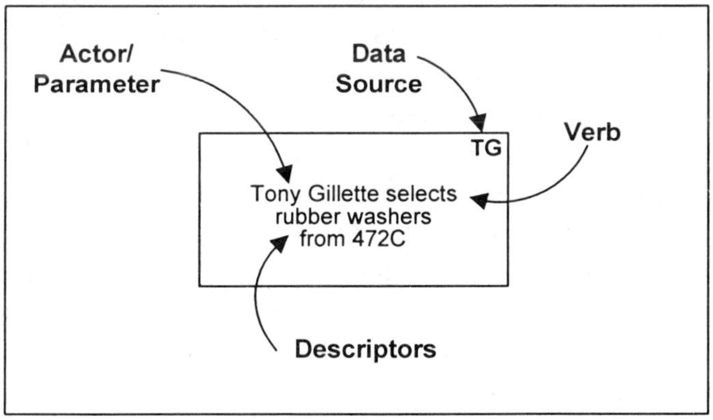

Figure 3-2 Abbreviated Building Block

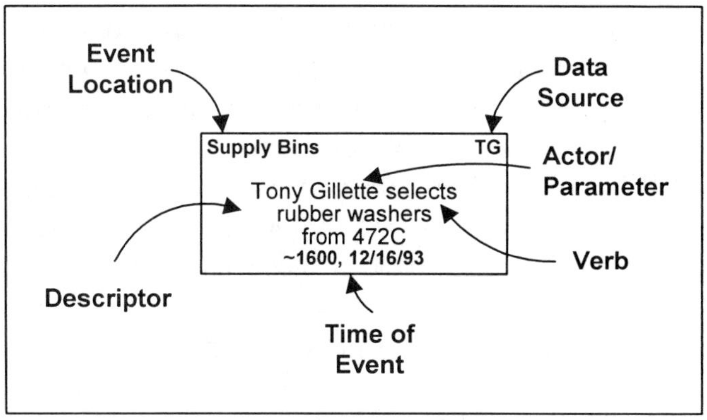

Figure 3-3 Comprehensive Building Block

Rectangles are used to illustrate *events and conditions* and diamonds are used for *questions*. Some investigators use one color for events and conditions and a different color for questions; in this case, using diamonds is not necessary. For *facts* we use a solid line and for *suppositions* we use a broken line. When using Post-It Notes, you do not need to put a solid rectangle around events that are facts. If only part of the data is supposition (such as "time"), then simply place a dashed line under that portion of the data on the building block. If all of the information on the building block is a supposition, underline all pieces of data.

Procedure

Follow these steps to prepare any building block:

1. **Think actions (and then conditions)**. A building block consists basically of one actor/parameter + one verb, accompanied by limiting descriptors. *Events/actions* typically cause accidents, and *conditions* typically contribute to accidents. Therefore, during investigations, keep listening and looking for actors whose actions initiated a change of state during an accident sequence.

2. **Always write the name of the actor/parameter before you write anything else**. You will be tracking actors/parameters. Actors can be people or things. For example, an actor can be Tony Gillette (a person), Valve V-23 (a thing), or temperature (a parameter). Give each person or thing a name,

then *use only that name thereafter*. Enter only *one specific* name on each building block. *Pronouns and plural nouns* are **poison words** for the analyst. The "actor" you name must be the "doer" of the action, rather than something or someone that was acted upon or had something done to it. Criteria for *condition* building blocks are not really necessary since we tend to naturally default to passive tense; however, it pays to be concise.

3. **Next, write down what the actor did or the state of the parameter, using past tense verbs with descriptors**. An actor's name must be followed by the action verb describing the action that initiated a change of state in someone or something during the incident. People actions include:

   - Sensory actions (like saw, heard, smelled, touched, or tasted — data acquired by the five senses)
   - Physical actions (like turned, reset, hit, pushed, pulled)
   - Mental actions (like decided, calculated, chose, concluded, thought)

   For things and parameters, actions should be descriptive of what they did (rose, struck, separated, scored, deposited, etc.) and not generic classifications such as failed, erred, etc.

   The verb should be qualified by additional phrases to achieve specificity (see Step 6 below). NEVER use two verbs in one building block!

4. **List the source of the data**. Every piece of evidence originated somewhere. Some data come from an analysis of the affected parts, debris, etc. Other evidence may come from computer files, dataloggers, logbooks, a person's testimony, etc. The investigator should list the source of the evidence on each building block. Developing a code for each source may be helpful if the source descriptions are long or there are numerous data sources. For evidence from witnesses, simply provide the initials of the person who supplied the information, even if the evidence pertains to another person. The listing of the data source will help correlate each building block to the supporting (detailed) data files.

5. **Enter times for the event/condition**. You will use the time an event or condition began and ended for placement on the CF chart and for truth testing. Therefore, the beginning and ending times should be acquired or estimated for each event or condition. The time each event *began* should be entered on the building block. If time is essential but is not precisely known, underline the time with a dashed line to show that the time is estimated, and investigate the time further. In many cases, simply the duration of an event or condition is important.

6. **Enter the location where the event/condition began/occurred**. Next, if physical motion or movement is involved, show where the actor was when his/her/its action began. This may require coordinates and reference(s) to points on sketches or maps.

7. **Enter descriptors limiting the action**. After entering the previous data, you are now ready to enter any descriptors needed to limit the action (quantitative measures, name of the person or thing that started to change state when acted upon, etc.). For example, in the event "Driver pressed on brake pedal with all her strength," the words following the verb 'pressed' are limiting descriptors used to make the event more specific. In the event "Aircraft descended 500 feet," *500 feet* is the limiting descriptor. These descriptors should be as specific as the data allow.

Recheck each building block before you use it. Before you use a building block in the subsequent events analysis steps (such as CF charting), recheck it to be sure you have recorded all of the available data correctly, cross-checking it against the original data if necessary. When the recheck is complete, you are ready to proceed to CF charting and other analysis methods.

Sometimes you will not be able to enter all of the data for every event. Make sure you obtain the information in Figure 3-2 and try to obtain all the information shown in Figure 3-3.

Optional Procedural Steps

One feature of CF charting is its flexibility. Investigators often find it useful to add special data about an event to a building block. Use whatever codes you wish, but make a note about each code used so you will use the data consistently when writing a report. For example, a witness may provide a photograph or sketch in addition to a statement. If so, you may wish to add a "P" or "Sk" to the source entry. Spatial references such as maps or sketches may be needed to track an actor's actions during an accident; you may wish to develop a "location" grid code tied to a map and "x,y" map coordinates or symbols. Using "C," "S," or "R" to indicate that the event was addressed by a code, standard, or regulation might be useful. Once recorded on the building block, coded entries will remind you of the issue when you use the building block in discussions, presentations, analysis, report writing, etc.

Application of the Building Block Procedures for the Two Primary Data Sources

One procedure can be used to document data derived directly from people during interviews. Another procedure can be used to document data from things or from people in noninterview situations.

A. Documenting Witness Testimony

Two types of data are available from a witness interview, simulation, or similar direct observation of witness actions:

- The actions of the witness
- The actions of other people or things the witness observed

Either type may be:

- Explicitly reported by a witness or
- Inferred from a witness' testimony

Data about witnesses' actions are usually acquired:

- In writing from written witness statements,
- Verbally and visually during witness interviews, or
- From documentation of witness actions, recorded on:
  - audio and/or videotape,
  - instrument charts,
  - operating logs,
  - license examinations,
  - etc.

Be alert for other data sources that might have observed and "recorded" the witness' actions.

1. **Written witness statements**. If the witness wrote an initial or follow-up statement, or if you use a written statement that you or another interviewer recorded, create the building blocks as follows:

   a. Underline everything the witness says she/he did (action words).
   b. Number the action words that imply that the witness did something. Be especially alert for comments suggesting that the witness moved, chose, concluded something, etc.
   c. Find and circle those specific actions by the witness that initiated a change of state:
      - in the witness or
      - in someone or something else
   d. Write the actor's name and each circled action on a self-stick note in the actor/action format. Each self-stick note becomes an actor/action phrase that constitutes an event building block (also note the source of the building block at this time).

2. **Witness interviews.** If you use personal or telephone witness interviews, do the following:

   a. Listen for and make note of the actions the witness took by writing down the action verbs and any additional descriptors you hear. As you listen, try to form a "mental movie" of the actions being described by a witness.
   b. During the interview, listen for descriptions or inferences of actions by other people or things. When you hear one, write down the actor's name and the action verb that accompanies the name.
   c. As soon as feasible, transfer your notes about events and conditions onto self-stick notes to form a building block. Do this before your memory fades.
   d. Although a detailed discussion of interviewing skills is outside the scope of this handbook, you may find it helpful to write your notes down in a narrative or outline format and offer them to the witness to correct and initial, signifying the witness' concurrence with what you recorded. This is a useful way to ensure that your information is correct.

At the conclusion of this effort, you should have a significant quantity of building blocks (probably with many redundant items) to use for analysis.

B. Documenting Events from Observations of Things

Although this documentation method is similar to that described above, acquiring "things" information upon which the building blocks are created is different. Things cannot tell you what they did or saw. The investigator has to "read" the information from the thing that provides it. This involves many special skills and a technical background on the things. Unless you are an expert, you may wish to obtain expert help in developing this information.

Once the things data are available, the general procedure for creating building blocks from things data is the same as described previously. However, (1) the actor will have a "thing" name rather than a "person" name and (2) the actions will be different (such as shifted, breached, punctured, actuated, separated, eroded, stopped, rotated, combined with, flowed, bridged), depending on the things involved.

Laying Out the Chart

Links Between Components

The basic elements of the CF chart are linked together by the relative positions of the building blocks. The arrangement serves to complete the chart by illustrating the relationships among elements. For summary definitions of the building blocks of the CF chart, see Table 3-1.

The CF Chart: Format

Traditionally, the format of CF charts has varied widely; however, the SOURCE methodology uses some general guidelines for developing CF charts. The underlying philosophy is that standardization will help ensure consistency and comparability (auditing) in event reporting within your organization. In addition, common guidelines will facilitate communication among personnel who routinely prepare incident reports and those who review these reports.

Development of the CF Chart

The guidelines for the CF chart format are listed in Table 3-2. They are not complex. The intent is only to provide some basic structure, not to inhibit investigators with many complex and cumbersome rules.

The first step in laying out the chart is to capture the general sequence of events and conditions that are known from the initial information provided from personnel, parts, and paper. You will want to arrange the building blocks on a CF chart as they are prepared.

Table 3-2 Guidelines for CF Chart Format

Primary and secondary events and conditions should be enclosed in rectangles	Primary Event/ Condition; Secondary Event/ Condition
Events or conditions not based upon valid factual evidence should be clearly identified by enclosing them in dashed rectangles, or by underlining the specific data that are suppositions	Presumptive Event; Presumptive Condition
The primary sequence of events should be depicted in a straight horizontal line with events joined by connecting arrows drawn in bold print. The loss event is usually at the end of the primary event line	Primary Event → Primary Event
Secondary events and conditions should be depicted on horizontal lines at different levels above and below the primary sequence of events. Secondary events should be joined to each other and to primary events by solid connecting arrows	Secondary Event ↓ Secondary Event ↓ Primary Event
Conditions should be connected to each other or events	Condition ↓ Event ↓ Condition
Events should be arranged chronologically from left to right and top to bottom; however, a sequence of secondary events/conditions may be on a different time line than the primary event/condition they connect to	1 → 2
Causal factors should be identified by lightning bolts. Items of note should be emphasized by asterisks	Causal Factor; * Item of Note

Create a chart background by taping together flip chart paper or using a large strip of wrapping paper. This will allow you to retain the completed chart in a file for later reference. (Alternately, you could take a photograph of the completed chart for use in the report and for archival purposes, or you could convert the chart to a hand- or computer-drawn flowchart.)

Start the chart by putting a *tentative* time line or time markers along the top or bottom of the chart. Use major events in the sequence such as an explosion, a release, an alarm activating, a shift change, the startup of the system, etc., as key time markers on the time line. Some of these events may be used to link the evidence collected from various sources together. For example, although the witnesses may have been located throughout the facility, their event time lines can be linked by the actuation of the evacuation horn since they were all able to hear it when it actuated. You are now ready to place each event building block onto the chart as you finish it. As the chart develops, you may find it necessary to expand the time line scale. Feel free to do so at any time. Using self-stick building blocks makes it easy to move the events around during the development of the chart.

The first building block you place on the chart should be the primary loss event or condition. If the scope of your investigation includes post-event analysis (such as analysis of the emergency evacuation and response efforts), this building block should be placed a third of the way from the right hand side of the chart (time progresses from left to right on the chart). Then, identify the event or condition that probably immediately precedes the loss event/condition. Keep working backward in time until the data are exhausted or you reach an **unknown**. When you arrive at an unknown, you will need to develop questions (further analysis needs) and place these on the chart at that location on the CF chart. These questions may be resolved directly by gathering more data.

If the data do not allow a direct answer to the question(s) and the gap prevents you from understanding what happened and why it happened, try using *fault tree analysis* to bridge the gaps in your logical event flows. Fault tree analysis is a *deductive* reasoning technique that helps you identify what data you should be seeking to verify your educated guesses.

If you do not have sufficient understanding of the physical/chemical process to allow construction of a fault tree, then your team may want to use an *inductive* technique, such as hazard and operability (HAZOP) analysis or failure modes and effects analysis (FMEA) to structure "brainstorming" of what could go wrong, which in turn will allow development of a more comprehensive fault tree. The fault tree can be constructed separately from the CF chart and should be used to guide further data collection to either eliminate or support a possible event path. If more than one possible path remains for the event that you are stuck at, even after more data collection, then it is wise to retain each path as a possible path to bridge the gap in the data.

Next, examine the events on the chart to verify relationships among the events. Look at the data provided by all data sources to determine event-condition relationships. Test each event against the preceding event and the following event to ensure that it is in its proper time and spatial sequence. This will help provide further order to the chart. Note that events that could have taken place months before the event may be a direct cause of an event that takes place far into the event sequence.

As these relationships are identified, the primary sequence of events will start to become evident. Initially, it may be difficult to determine whether an event is primary or secondary. Each building block should be added to the "skeleton" chart as it is uncovered. As additional facts are uncovered, a more complete picture of the occurrence will emerge, and it will become easier to determine whether or not an event is directly related to the loss event. Modifications can be made to the chart as appropriate.

Presumptive events and conditions should be clearly identified as such by enclosing them with dashed lines. Every effort should be made to substantiate presumptive events and conditions with factual evidence. Allowance of presumptive components on the chart should not provide an excuse for a less than thorough investigation.

See Figure 3-4 for an example of a CF chart that contains events, conditions, suppositions, and questions (or possibilities that have not yet been eliminated).

Quality Control Checks

Check your building blocks for certain *poison* words. This step could be done with each building block as it is developed.

- Poison words like "*and*" and "*or*" indicate that you are covering more than one event or condition in a building block.
- "*Was*" and "*were*" reflect the passive voice, which indicates that you may not have named the right actor. Passive voice is acceptable, however, for stating *conditions*.
- Words ending in "...*ly*" are usually poison, because they are merely disguised, implicit investigator value judgments with no criteria or evidence to justify their conclusion.

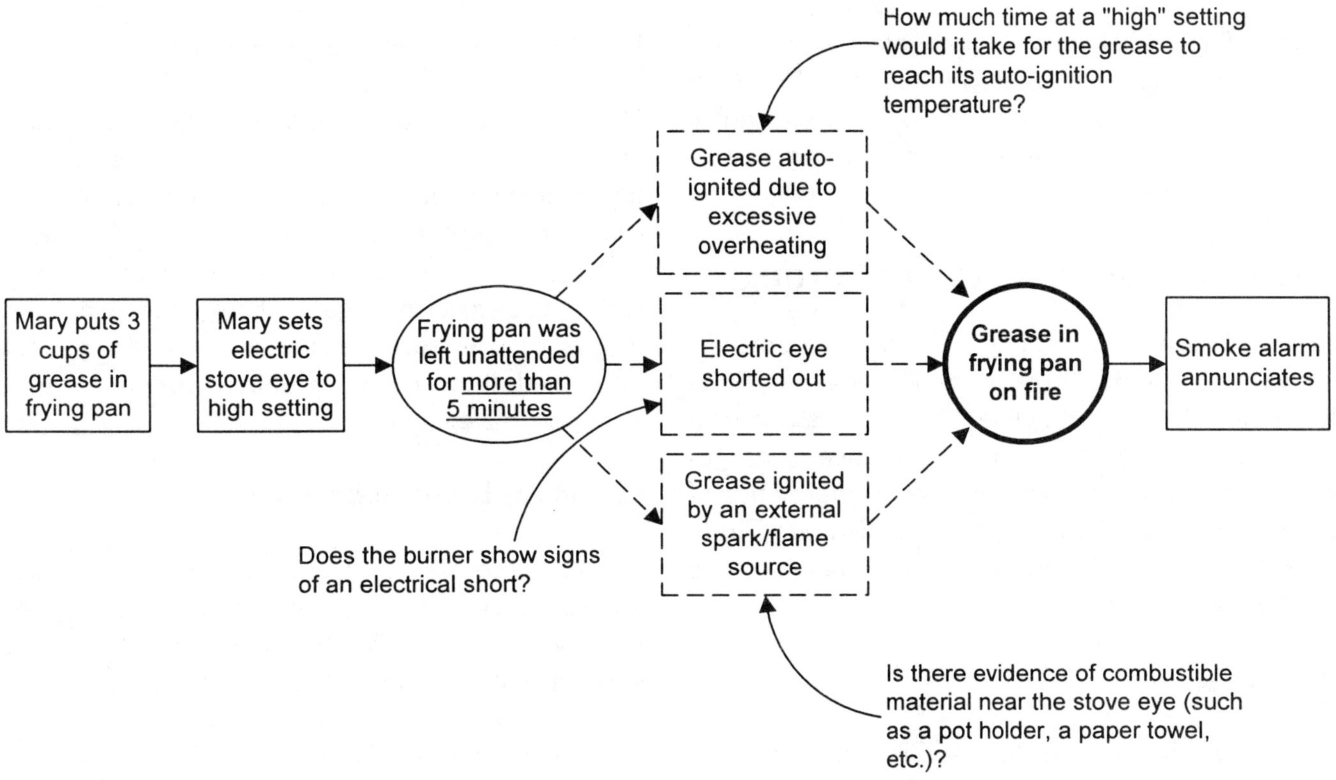

Figure 3-4 Incomplete CF chart for a Grease Fire on an Electric Stove

- Other poison words are words that describe a steady state, rather than an event or condition.

As you gain experience, you will begin to spot your own signals that a building block is really not an event at all, but rather is a conclusion, supposition, or other unacceptable entry. Summary statements (which are also conclusions, but are clearly based on data) are acceptable (and useful) on a CF chart.

After a chart is completed, have someone who does not know what happened review your chart for the logical flow and sequential ordering of the event. This process often generates either a need for more data or better event flows.

Causal Factor Identification

After the CF chart is complete, the investigators are in an excellent position to identify factors that were major contributors to the incident. Determination of these causal factors requires judgment on the part of the investigators. Causal factors are those human errors and equipment failures that, if eliminated, would have prevented the event or reduced its consequences. Furthermore, presumptive events and conditions may be identified as presumptive causal factors.

It is important to remember that most incidents do not have a single cause. Usually a number of factors contribute to an incident. The evaluation should not stop after the first causal factor is discovered. The investigators should continue until all major contributors to the incident have been identified. Each of the causal factors should be marked on the chart using a lightning bolt symbol.

Practical Application of CF Charting

The instructions for preparing a CF chart presented in this handbook should serve as basic guidelines. Strict adherence to these guidelines is not necessary. The combined experience of many occurrence investigators has led to the identification of several "rules of thumb" to follow when developing a CF chart. These suggestions, applied judiciously, should help achieve high-quality investigations.

Start early.

The investigators should start the CF chart as soon as they begin to collect facts about the incident. They should construct a "working chart," starting with the loss event and working backward in time. This will be only a skeleton of the final product, but it will serve to ensure that valuable information and questions are not forgotten or lost during the investigation. Care should be taken, however, to avoid locking the investigating team into a preconceived scenario.

Follow format guidelines.

The investigators should use the guidelines for formating described in this document. This will help them get started and stay on track as they reconstruct the events and conditions surrounding the incident. Proper perspective should be maintained in applying the guidelines. They are intended to assist you in simple application of the investigative tool. They are not hard and fast rules that must be applied without exception. They have grown out of experience and fit well for most situations. If the investigators believe that they have a truly unique situation and need to deviate from the guidelines, then they should feel free to do so.

Proceed logically with available data.

Naturally, events and conditions are not going to emerge in the sequence in which they occurred during the incident. Initially, the CF chart will have many unresolved gaps and questions. The job of the investigating team is to probe deeply enough to get the facts needed to fill these gaps. Use "gap bridging" techniques, such as fault trees, to deduce what might have happened, then search for data to support or eliminate possible event paths.

Use an easily updated format.

As more information is learned about the incident, the working CF chart will need to be updated. It is of extreme importance to choose a format that can be easily modified; otherwise, the chart will become too cumbersome to be of any value. In the past, investigators have attempted to redraw the chart repeatedly. This approach is both time consuming and frustrating for those involved in the investigation, even when using computer software specifically created for this purpose (the monitor is too small to allow viewing the entire chart). The technique that has proven most effective involves the use of self-stick notes (e.g., Post-it Notes) and a large sheet of paper (e.g., flip chart paper or wrapping paper). A single event or condition is written on each self-stick note and affixed to the paper. This technique has the advantages of being simple, not requiring any special tools or software, and being low-tech. This allows the investigators to focus on the investigation instead of the method of updating the chart. As a more complete picture of the occurrence emerges, the self-stick notes can be added, deleted, or rearranged. Using the large sheet of paper as a base allows the investigators to take the chart with them if they need to move among conference rooms, offices, or locations involved in the event. Once the working chart has been completed, a final version can be drawn for inclusion in the occurrence report. Experienced occurrence investigators have discovered that the chart's most useful features are that it illuminates "gaps" in knowledge, points to areas for further inquiry, and makes report writing relatively straightforward.

Use other investigative techniques when appropriate.

CF charting provides a way for investigators to organize the data collected during the incident investigation. Numerous techniques exist for collecting the data that go into the chart. Some of these are presented in documents listed in the References section of this handbook. The more skill the investigators have in collecting event data, the better the CF chart. The investigators should make every effort to build their knowledge of investigative techniques.

Select the appropriate scope for the CF chart.

One of the first things to consider when creating the CF chart is the scope of the investigation. It is necessary to decide, on a case-by-case basis, upon the appropriate depth and sequence length of the CF chart. Unless the scope is defined early, much time will likely be wasted collecting information that will later be discarded because it falls outside the boundaries of the final investigation. For example, many investigations are limited to pre-event events and exclude post-event items (such as emergency response).

Provide an executive summary of the CF chart.

Condense the working CF chart into an executive summary chart for publication in the event report. The working chart will contain much detail, so it is of greatest value in guiding the investigation. However, for the event report, the primary purpose of the CF chart is to provide a concise, easy-to-follow representation of the event sequence for the report readers.

Advantages of CF Charting

The benefits of CF charting for event investigation are numerous. Several of the most obvious advantages are listed below.

Provides organization of data.

The chart provides a way to organize the data gathered during the event investigation. Often, important data are lost or forgotten as the investigation progresses. If the investigation is being conducted by a team, different investigators may collect different pieces of information. The CF chart helps ensure that everyone involved has the benefit of the group's total knowledge.

Guides the investigation.

The technique is excellent for group investigations. The CF chart provides a common reference for everyone involved. While inexperienced investigators can use the method to structure their investigations, experienced investigators can use the chart as a way to avoid drawing conclusions before they have all of the relevant facts. CF charting forces investigators to think about causal factors, one at a time, instead of considering the occurrence in global terms.

Allows validation of the incident sequence.

A CF chart provides a good reference during interviews for those directly involved in an event. The investigators can ask interviewees if the chart is correct. Interviewees have a graphic representation of what the investigators think happened during the event. They can easily point out discrepancies. Note that we recommend **not** showing the chart to interviewees until the follow-up interview phase.

Allows identification of causal factors.

Many times event investigators are tempted to think of an event in global terms. They ask themselves what they can do to "fix the problem." Using the global approach, we often address only parts of the problem. A CF chart allows us to see the entire event, broken down into its components. Each part can be assessed separately, and solutions can be recommended for individual causal factors. This lessens the probability that some important contributor to the event will be overlooked.

Simplifies organization of event report.

Generally, a graphic representation of an event is more easily interpreted than a narrative representation. Readers of an event report can glance at a CF chart and quickly familiarize themselves with the event. Gaps in logic that might not be so visible in a narrative report are far more apparent when presented in chart form. CF charting has proven to be a clear and concise aid for report readers whose goal it is to understand the causes of the event.

Summary

CF charting (together with continued data gathering) is the second major step in the root cause analysis process. The technique is simply a tool designed to help event investigators describe the events leading up to and following an event, as well as the conditions surrounding these events. The technique provides a structured approach to collecting and analyzing the facts pertaining to an event. It is up to the investigators to decide how detailed the investigation, and thus the CF chart, should be.

The charting technique, in and of itself, does not ensure an adequate event investigation. The investigation team must be knowledgeable about the processes, facilities, and personnel involved in the event. They must know the right questions to ask, who to ask, and how to ask. Finally, they must be willing to probe to the levels necessary to determine WHAT happened during the occurrence, to describe HOW it happened, and to understand WHY. (The following chapter on root cause identification explains how to consistently and thoroughly understand WHY.)

CHAPTER 4

ROOT CAUSE IDENTIFICATION

Once the investigator has created a CF chart describing the event and filling in all the gaps possible, the next step in the SOURCE root cause analysis process is to determine the root cause(s) for each causal factor (or item of note) identified in the chart. *Root causes are the most basic causes that can reasonably be identified, which management has control to fix and for which effective recommendations for preventing recurrence can be generated.* Root cause identification is simply a process to help the investigator determine the underlying problems (i.e., root causes) associated with each causal factor. As defined earlier in this handbook, *causal factors are human errors and equipment failures, which if eliminated, would have either prevented the event or reduced its severity*. A causal factor, as identified in the CF chart, is a description of *WHAT* happened to cause the incident or *HOW* it happened. Before recommending workable preventive measures, the investigator must know *WHY* the causal factor occurred. Root cause identification helps the investigator examine, in a systematic way, possible reasons for the causal factor.

The Root Cause Identification Process

For simplicity, a map format was chosen for use in root cause identification. Investigators use the Root Cause Map (RCM) to structure their reasoning process. Consistency is ensured across all investigations by using the same process for categorizing causal factors. A copy of the map is presented as Appendix B to this handbook. Starting at the top of the map, the investigator identifies the root causes (and higher categories of causes) for each causal factor identified in the CF chart by working down through the map as far as known information will allow. For each causal factor, an investigator determines which top level node is applicable. Based on this decision, the investigator moves down to the next level and selects another applicable node, keeping in mind that only lower level nodes branching from the node chosen on the previous level can be considered for identifying the root causes of a causal factor. Paths through the RCM flow only in a downward direction. By following the segments of the map, nodes that do not apply to a given causal factor are not considered, saving considerable time and effort during the identification process.

Figure 4-1 shows the map format and demonstrates how to follow a path from the top of the map to the lower levels. The arrows show examples of possible paths. When using the RCM, the investigator always starts at the top of the map with a given causal factor and proceeds down through the map as far as possible, given the information available. Movement through the map is always from the top down. If information is not available to answer questions at the lowest level of the map, the investigator can stop at a higher level. For example, in Figure 4-1, the path at the far left stops short of the lowest level.

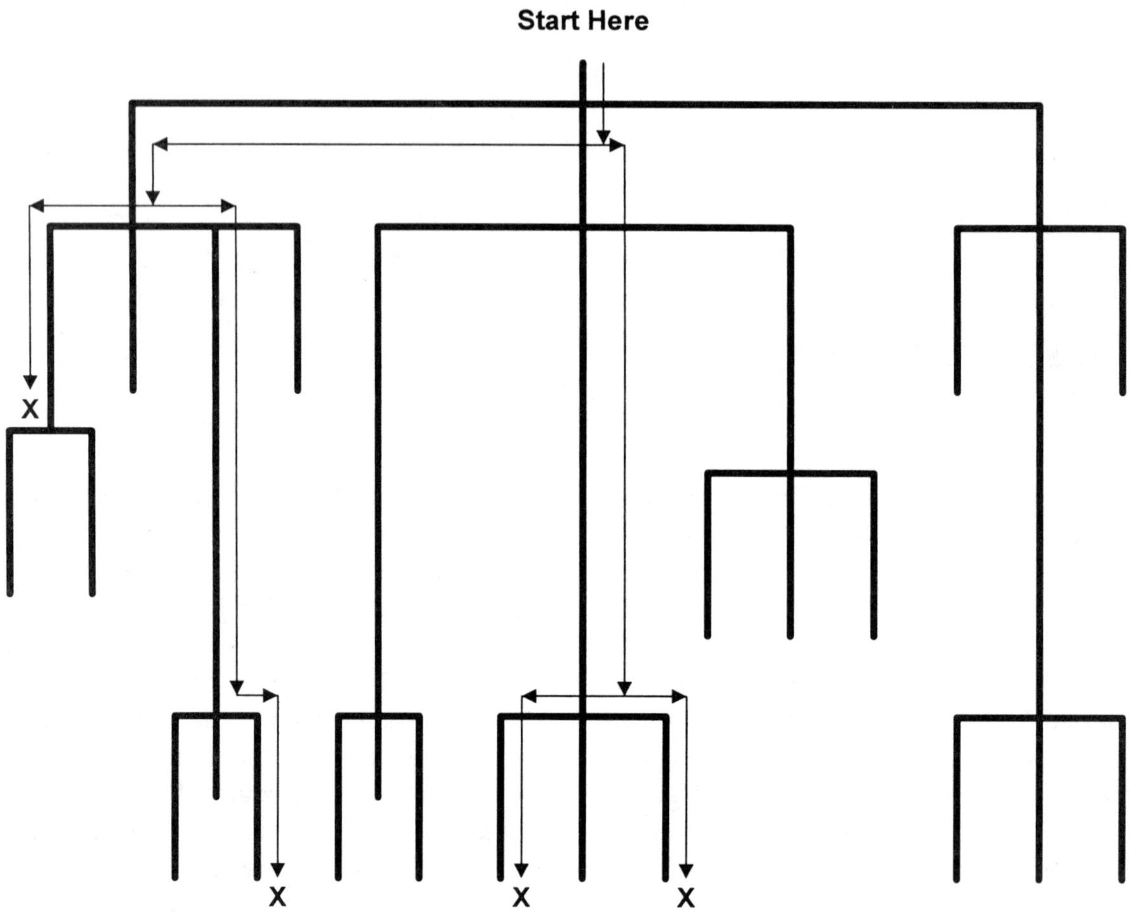

Figure 4-1 Use of the Map Format

The Root Cause Map

The Root Cause Map, the major tool used in root cause identification, is a decision diagram divided into many different nodes. At first glance, the RCM appears to be unbalanced. The left side contains far fewer nodes than the right side. Basically, the map is divided into two major parts. Nodes on the left side of the map are used to identify and categorize causal factors associated with equipment failure. Nodes on the right side of the map are used to identify and categorize causal factors related to personnel error. This division is illustrated in Figure 4-2.

Notice that the two sides of the map are not mutually exclusive. All failures, with the exception of other difficulties, can eventually be traced back to some type of human error. Even in the case of a lightning strike or flooding, it might be argued that the system designers did not provide adequate protective measures. In any event, equipment problems can often be traced back to mistakes made by personnel. For example, a pump malfunction may be the result of a maintenance mechanic's failure to correctly follow the required procedure. To deal with scenarios such as these, the two sides of the map intersect at a number of the second level (B level) nodes. This allows categorization from the equipment side of the map to extend over to the personnel side and vice-versa.

Segments of the Root Cause Map

As shown in Figure 4-3 (and the full-size map included with this handbook), the Root Cause Map has been divided into 11 segments. The segments have been color coded so that they are easily distinguishable. Each segment is made up of related nodes. For example, all nodes associated with communications in-

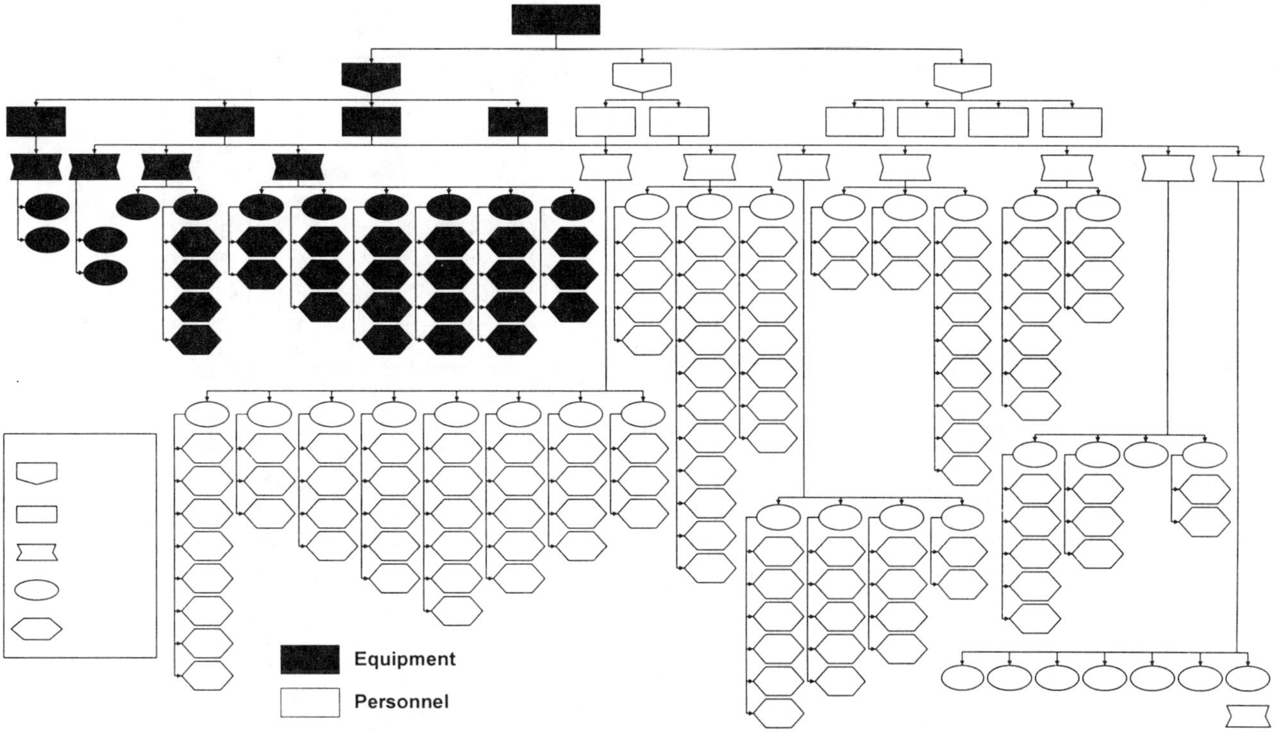

Figure 4-2 Equipment and Personnel Sides of the Root Cause Map

adequacies are located together in a single segment. All nodes related to training difficulties are grouped together in another segment.

Detailed Descriptions of Each Node

Detailed descriptions of each node are presented in Appendix A of this handbook. Typical issues addressed by each node are presented, usually in the form of questions. These questions are intended to help the investigator determine whether or not a node is appropriate for identifying and categorizing a particular causal factor. In addition to the typical issues, one or more examples of the types of events coded under this node are presented. Finally, a set of typical recommendations is provided to assist investigators in generating effective recommendations.

Levels of the Root Cause Map

In addition to dividing the RCM into segments, the map has also been divided into five major levels (see Table 4-1). Each level on the map corresponds to a particular class of nodes. When determining the root cause of a causal factor, Level A nodes require the investigator to make only broad distinctions. Level E nodes require that very specific questions be answered. The nodes on each level are coded by shape to help the investigator differentiate between levels. Table 4-1 provides a description of the different levels of the map.

This level is the most general level of the map. For each causal factor, the investigator is first asked to make broad distinctions concerning the type of difficulty involved. The investigator may initially determine that a particular causal factor involved an *Equipment Difficulty* or a *Personnel Difficulty*. Based on answers to these general questions, the investigator branches down to more specific levels of the map.

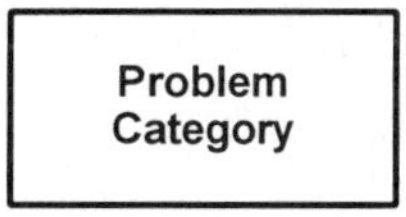

This level describes the type of problem that is being addressed. Examples of this level of nodes include *Equipment Design Problem* and *Company Employee*.

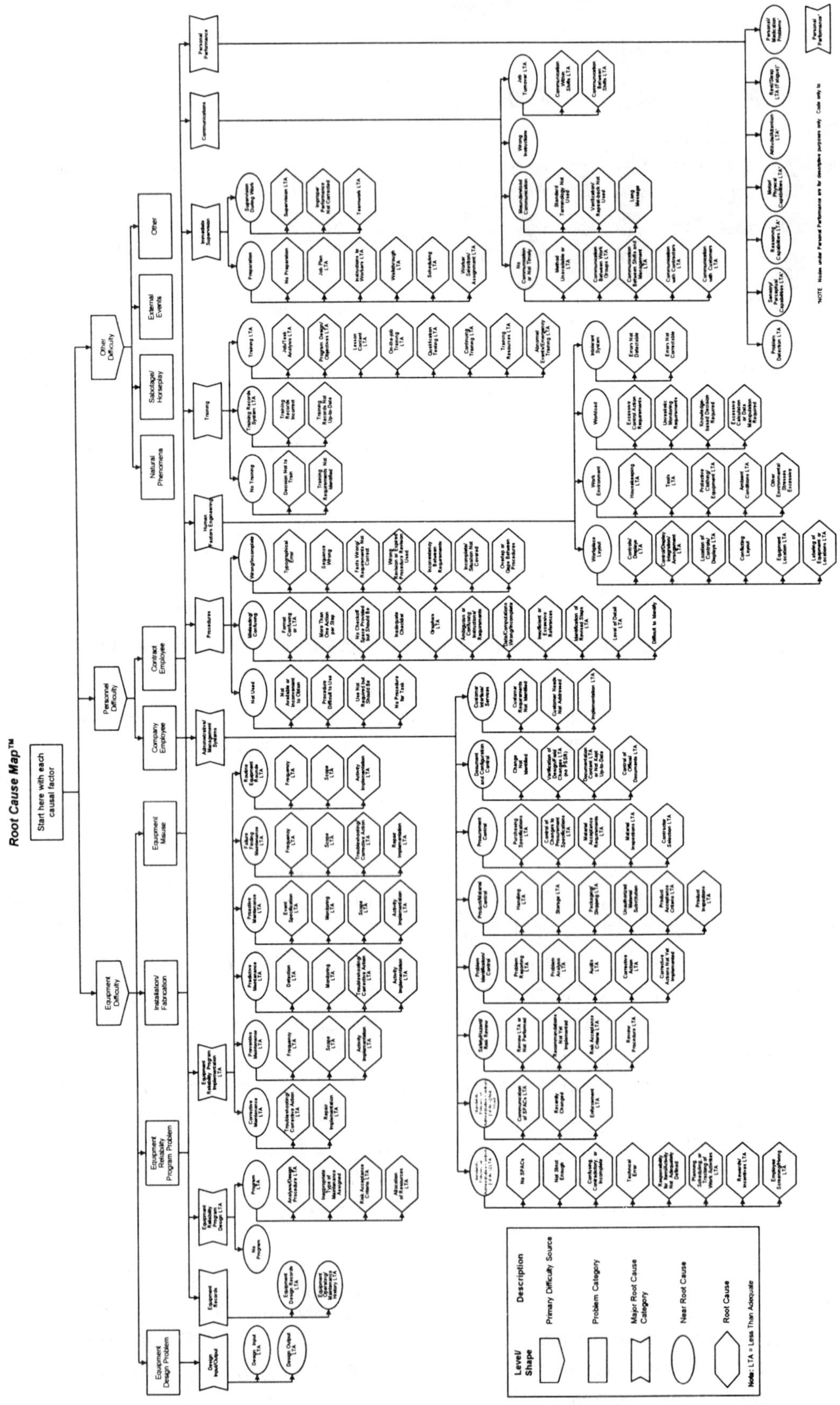

Figure 4-3 Root Cause Map

Table 4-1 Levels of the Root Cause Map

Level/ Shape	Description	Examples
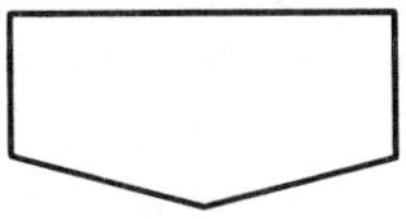	Primary Diffficulty Source	• Equipment Difficulty • Personnel Difficulty • Other Difficulty
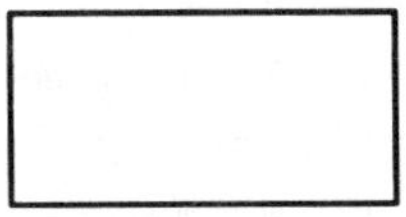	Problem Category	• Equipment Design • Installation/Fabrication • Natural Phenomena
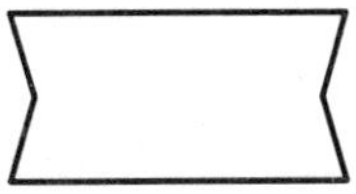	Root Cause Category	• Design Input/Output • Equipment Reliability Program Implementation • Human Factors Engineering
	Near Root Cause	• Preventive Maintenance Less Than Adequate • No Training • Procedures Misleading/Confusing
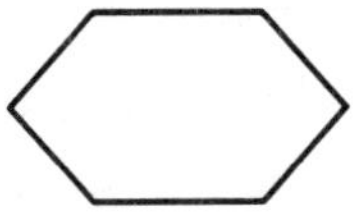	Root Cause	• More Than One Action per Step • Inappropriate Type of Maintenance Assigned • Training Records Incorrect

When the investigator reaches this level for a causal factor, the root cause categories must be determined. This level contains 11 categories; examples include *Design Input/Output*, *Training*, *Procedures*, *Administrative/Management Systems*, and *Human Factors Engineering*. Using this level, the investigator begins to become more specific about the nature of the causal factor; this is an essential categorization step in any root cause analysis.

The investigator now moves to the near root cause level of the RCM, which includes subdivisions of the major root cause categories. Notice that most of the nodes under the *Personal Performance* root cause category are shaded. The shaded nodes are intended to provide guidance concerning the types of issues that should be considered personal performance problems. Because of the sensitivity of personal performance issues, the event report should stop at and reference only *Personal Performance*. The reason for the personal performance problem should be treated confidentially through other channels (e.g., medical department, disciplinary system).

Finally, the bottom level lists a detailed set of root causes for each root cause category. This level, the lowest level of the map, requires that the investigator answer very detailed questions about the causal factor. Although we cannot predict all the questions that might pertain to a root cause node, Appendix A contains several examples to help the investigators consistently choose the best fit. Examples of nodes located at the lowest level of the map include *Wrong Revision or Expired Procedure Used*, *Packaging/Shipping Less Than Adequate (LTA)*, or *Labeling LTA*. The goal of root cause identification is to allow the investigator to be as specific as possible about the underlying reasons for a given causal factor. If possible, the investigator should attempt to reach the bottom level nodes.

An investigator could go beyond (below) the root cause level, but in most cases this will identify issues that cannot be addressed effectively by recommendations from investigation teams. It is probably best to stop at the root causes on the map, unless the data from many investigations indicate recurrence of a particular root cause.

Multiple Identification

Usually there are multiple root causes for a causal factor. This is referred to as multiple identification. Multiple identification is used whenever more than one root cause is responsible for the occurrence of a causal factor. Multiple coding is appropriate in many circumstances; however, if more than three root causes are indicated, the causal factor is probably stated too broadly and should be divided into two or three more specific causal factors. For example, a mechanic fails to properly perform corrective maintenance on a pump. Part of the reason for the improper repair is an out-of-date procedure. A different motor had been installed in the system and the procedure was not updated to reflect this change. This event can be coded under (1) *Equipment Reliability Program Implementation LTA*, *Corrective Maintenance LTA*, and *Repair LTA* (the repair was not performed correctly), (2) *Procedures*, *Wrong/Incomplete*, and *Facts Wrong/Requirements Not Current* (the procedure was not updated), and (3) *Administrative/Management Systems*, *Document and Configuration Control*, and *Documentation Not Kept Up to Date* (the system for updating documents failed).

A Simple Example: The Pump Failure

The best way to explain the mechanics of root cause identification is through a simple example. Consider the following event:

During the operation of a chemical process, the operator on duty observed the flow rate for one of the streams decrease rapidly and fall to zero. The operator halted feed to the process and shut down the operation per procedures. Investigation into the cause of the event revealed that the bearings in the pump seized, causing the pump to fail. It was determined that the bearings seized because they were improperly installed during routine maintenance. Therefore, the causal factor identified in the CF chart was "improper installation of the bearings during maintenance." When questioned, the maintenance mechanic stated that he had followed the written procedures for the task. When the procedure was examined, it was found to be outdated. A more recent revision should have been used, and its use might have prevented the bearing problem.

Starting at the top of the RCM, the investigator concluded that the source of the difficulty was an equipment problem. Therefore, the first node coded is *Equipment Difficulty*. Next, the problem category is determined to be *Equipment Reliability Program Problem*. Because the wrong version of the procedure was used, the next nodes to be identified would be *Procedures*, *Wrong/Incomplete*, and finally, *Wrong Revision or Expired Procedure Revision Used*. In addition, the procedure administration system failed to provide the proper revision of the procedure to maintenance personnel. Therefore, this causal factor could also be determined to have *Document and Configuration Control* as a near root cause and *Control of Official Documents Less Than Adequate* as a root cause, under the category of *Administrative/Management Systems*. Table 4-2 illustrates the paths followed through the Root Cause Map.

Presentation of Results

For the incident report, each causal factor and item of note should be presented using a three-column table such as that presented in Figure 4-4. As discussed in Chapter 3, the first column is used to describe the causal factor or item of note. The paths through the RCM used to categorize a particular causal factor are presented in the second column. The entire path through the map should be listed with the top level node listed first and the bottom level nodes listed last. If multiple root causes are identified for the causal fac-

Table 4-2 The Pump Failure: Paths Coded Through the Root Cause Map

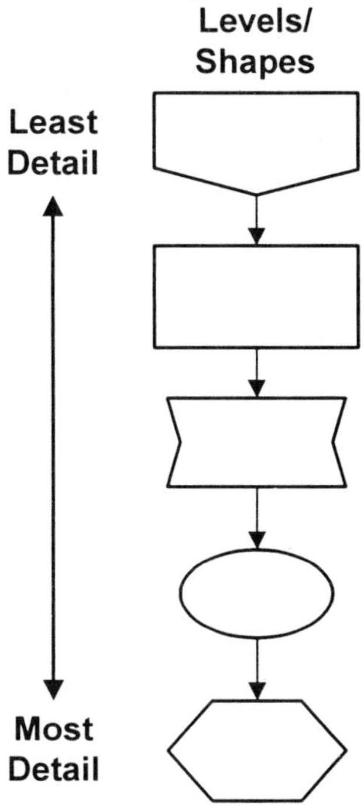

Examples	
• Equipment Difficulty	• Personnel Difficulty
• Equipment Reliability Program Problem	• Company Employee
• Procedures	• Administrative/ Management Systems
• Wrong/Incomplete	• Document and Configuration Control
• Wrong Revision or Expired Procedure Revision Used	• Control of Official Documents Less Than Adequate

tor, then all paths through the map should be shown. A solid line in the middle column should be used to separate the information about different paths.

Summary

Root causes are identified for each causal factor. Root causes are the most basic causes that can reasonably be identified, that management has control to fix, and for which effective recommendations for preventing recurrence can be generated. The Root Cause Map is used to structure the identification process. The investigator starts at the top of the map and moves down through the paths on the map to identify root causes. Multiple root causes are identified for each causal factor. The results of the root cause identification process are presented using a three-column table.

Title of Event Investigation Report		
Causal Factor #1	**Paths Through Root Cause Map**	**Recommendations**
In the first space in column one, repeat the causal factor as it appears in the causal factor chart. **BACKGROUND:** In this space, provide a brief background that explains the causal factor listed, including underlying reasons why the causal factor occurred or was allowed to exist. This space could also be used to elaborate on the causal factor, if the causal factor statement is terse.	• Primary difficulty source • Problem category • Root cause category • Near root cause • Root cause In this space, list the paths through the map to a root cause.	In this column, list (or reference) any recommendations (or decision not to make a recommendation) to address each root cause.
	• Primary difficulty source • Problem category • Root cause category • Near root cause • Root cause List the paths for any other root cause(s).	List or reference recommendations to address the other root causes associated with this causal factor

Figure 4-4 Root Cause Summary Table: Presentation of Paths Through the Root Cause Map

CHAPTER 5

RECOMMENDATION GENERATION AND IMPLEMENTATION

Perhaps the most significant aspect of root cause analysis is the final step. Following the identification of root cause(s) for a particular causal factor, recommendations for preventing its recurrence must be generated. The identification of effective corrective actions is addressed explicitly in the definition of root causes. Root causes are defined as *the most basic causes that can reasonably be identified, which management has control to fix and for which effective recommendations for preventing recurrence can be generated.* The emphasis is on correcting the problem so that it will not be repeated. The following criteria for ensuring the viability of corrective actions are suggested.

- Will these corrective actions prevent recurrence of the condition or event?
- Is the corrective action within the capability of the organization to implement?
- Are the recommendations directly related to the root causes?
- Can we ensure that implementation of the recommendation will not introduce unacceptable risks?

The corrective actions developed should address not only the specific circumstances of the event that occurred, but also system improvements aimed at the incident's root causes. They should address options for reducing the frequency, minimizing the personnel exposures, and/or lessening the consequences of one or more of the root causes.

In general, three types of recommendations should be generated for each root cause:

- Correct the specific problem
- Correct similar existing problems
- Correct the system that created the problems

For example, if a procedure error is discovered, the following three recommendations may be generated:

- Correct the specific procedure that caused the problem under investigation
- Correct other procedures that have this same type of problem
- Correct the procedure-generation process to prevent the creation of more problems

A suggested format for recommendations is to provide a general objective to be accomplished, followed by a specific example of how it could be accomplished. This will clearly state the intentions and provide a specific method that can be used to accomplish the goal. If management wants to implement this suggested solution they may. However, they may also have a better, more effective way to meet the objective.

For example, a recommendation may state:

- Provide a means for operators to detect slow changes in tank levels. For example, a strip chart recorder that shows trends over 8 hours could be provided in the control room.

Management may decide to provide tank levels on the plant process computer so that operators can change the time scale to whatever they need, from minutes to days.

The recommendations should seek to make improvements in management systems and/or inherent safety by:

- Reducing the inventories of hazards
- Making substitutions for hazardous chemicals or situations
- Increasing the number of events required to generate an incident

Finally, ensure that the recommendations are based on the conclusions from the data analysis results. The facts discovered during the investigation should lead to causal factors and root causes, which, in turn, should lead to the recommendations. Facts are relevant if they lead to a recommendation. Facts that are not required to understand the sequence of events or that do not support a recommendation should not be included in the investigation report. Likewise, all of the recommendations should be derived from the circumstances of the event and its root causes.

Checklist for Developing and Implementing Corrective Actions

In developing and implementing corrective actions, consideration of the following questions can help ensure that the criteria listed above are met:

- ❑ Is there at least one corrective action associated with each root cause?
- ❑ Does the corrective action specifically address the root cause?
- ❑ Will the corrective action cause detrimental effects?
- ❑ What are the consequences of implementing the corrective action?
- ❑ What are the consequences of not implementing the corrective action?
- ❑ What is the cost of implementing the corrective action?
- ❑ Will training be required as part of implementing the corrective action?
- ❑ In what time frame can the corrective action reasonably be implemented?
- ❑ What resources are required for successful development of the corrective action?
- ❑ What resources are required for successful implementation and continued effectiveness of the corrective action?
- ❑ What impact will the development and implementation of the corrective action have on other work groups?
- ❑ Is implementation of the corrective action measurable?

Presentation of Results

For the event report, each causal factor and item of note should be presented using a three-column table such as that presented in Figure 4-4. As discussed in previous chapters, the first column is used to describe the causal factor or item of note. The paths through the Root Cause Map are presented in the second column. The third column should be used to present recommendations for corrective action. If multiple identification has been used to categorize a causal factor or item of note, then at least one recommendation for each root cause should be presented. The recommendation(s) for preventing recurrence of the causal factor should be presented adjacent to the corresponding root cause. The person(s) or organization(s) responsible for implementing a given recommendation should be listed in parentheses immediately after that recommendation.

Summary

Recommendation generation is the final step in the SOURCE root cause analysis process. When this step is finished, the documentation of the root cause analysis is complete. The final product will be a CF chart and a root cause summary table for each causal factor or item of note identified during the root cause analysis process. In generating recommendations, it is important for the investigator to be confident that (1) the corrective actions will prevent recurrence of the event or condition, (2) each corrective action is within the capability of the organization to implement, (3) the corrective actions allow the organization to meet its primary objective, and (4) all assumed risks have been considered (and stated, if appropriate).

CHAPTER 6

REFERENCES

American Institute of Chemical Engineers, *Center for Chemical Process Safety Guidelines for Investigating Chemical Process Incidents*, New York, NY, 1992.

Department of Energy, *Accident/Incident Investigation Manual, 2nd ed.,* DOE/SSDC 76-45/27.

Department of Energy, *Events and Causal Factors Charting,* DOE/SSDC 76-45/14, 1985.

Department of Energy, *Root Cause Analysis Handbook*, WSRC-IM-91-3, 1991 (and earlier versions).

E.I. duPont de Nemours & Co., Inc., *User's Guide for Reactor Incident Root Cause Coding Tree,* Rev. 5, DPST-87-209, Savannah River Laboratory, Aiken, SC, 1986.

Ferry, Ted S., *Modern Accident Investigation and Analysis, 2nd ed.,* John Wiley and Sons, New York, NY, 1988.

Occupational Safety and Health Administration Accident Investigation Course, Office of Training and Education, 1993.

ROOT CAUSE MAP™ NODE DESCRIPTIONS

This appendix is designed to help you use the Root Cause Map in a consistent manner. It contains a list of typical issues, typical recommendations, and examples for each node on the map. The appendix covers the nodes on the map from left to right and from top to bottom. To help you find information on a specific node, two node listings are provided.

The first list is organized by areas of the map. This listing is useful if you are interested in examining a particular portion of the map.

The second list provides an alphabetical listing of the nodes. This list can be used to locate information concerning a specific node.

A third index is actually contained on the Root Cause Map in Appendix B. On the map contained in Appendix B, each node has a number in the graphic following the text. The number in the graphic indicates the page number in Appendix A that corresponds to that node. By using this version of the map, a user can directly determine the appropriate page in the handbook without the need for a separate index.

Index of Root Cause Map Nodes by Area

Primary Difficulty Source **1**
Equipment Difficulty 2
Personnel Difficulty 3
Other Difficulty 4

Problem Category
Equipment Design Problem 5
Equipment Reliability Program Problem 6
Installation/Fabrication 7
Equipment Misuse 8
Company Employee 9
Contract Employee 10
Natural Phenomena 11
Sabotage/Horseplay 12
External Events 13
Other 14

Major Root Cause Category
Design Input/Output 15
Equipment Records 19
Equipment Reliability Program Design LTA 21
Equipment Reliability Program Implementation LTA 28
Administrative/Management Systems 55
Procedures 111
Human Factors Engineering 138
Training 163
Immediate Supervision 180
Communications 192
Personal Performance 208

Design Input/Output **15**
Design Input LTA 16
Design Output LTA 17

Equipment Records **18**
Equipment Design Records LTA 19
Equipment Operating/Maintenance History LTA 20

Equipment Reliability Program Design LTA **21**
No Program 22
Program LTA 23
Analysis/Design Procedure LTA 24

Index of Root Cause Map Nodes by Area (continued)

Inappropriate Type of Maintenance Assigned 25
Risk Acceptance Criteria LTA 26
Allocation of Resources LTA 27

Equipment Reliability Program Implementation LTA 28
Corrective Maintenance LTA 29
Troubleshooting/Corrective Action LTA 30
Repair Implementation LTA 31
Preventive Maintenance LTA 32
Frequency LTA 33
Scope LTA 34
Activity Implementation LTA 35
Predictive Maintenance LTA 36
Detection LTA 37
Monitoring LTA 38
Troubleshooting/Corrective Action LTA 39
Activity Implementation LTA 40
Proactive Maintenance LTA 41
Event Specification LTA 42
Monitoring LTA 43
Scope LTA 44
Activity Implementation LTA 45
Failure Finding Maintenance LTA 46
Frequency LTA 47
Scope LTA 48
Troubleshooting/Corrective Action LTA 49
Repair Implementation LTA 50
Routine Equipment Rounds LTA 51
Frequency LTA 52
Scope LTA 53
Activity Implementation LTA 54

Administrative/Management Systems 55
Standards, Policies, or Administrative Controls (SPACs) LTA 57
No SPACs 59
Not Strict Enough 60
Confusing, Contradictory, or Incomplete 61
Technical Error 62
Responsibility for Item/Activity Not Adequately Defined 63
Planning, Scheduling, or Tracking of Work Activities LTA 64
Rewards/Incentives LTA 65
Employee Screening/Hiring LTA 66

Index of Root Cause Map Nodes by Area (continued)

Standards, Policies, or Administrative Controls (SPACs) Not Used 67
Communication of SPACs LTA 69
Recently Changed 70
Enforcement LTA 71
Safety/Hazard Review 72
Review LTA or Not Performed 74
Recommendations Not Yet Implemented 75
Risk Acceptance Criteria LTA 76
Review Procedure LTA 77
Problem Identification/Control 78
Problem Reporting LTA 80
Problem Analysis LTA 81
Audits LTA 82
Corrective Action LTA 83
Corrective Actions Not Yet Implemented 84
Product/Material Control 85
Handling LTA 87
Storage LTA 88
Packaging/Shipping LTA 89
Unauthorized Material Substitution 90
Product Acceptance Criteria LTA 91
Product Inspections LTA 92
Procurement Control 93
Purchasing Specifications LTA 95
Control of Changes to Procurement Specifications LTA 96
Material Acceptance Requirements LTA 97
Material Inspections LTA 98
Contractor Selection LTA 99
Document and Configuration Control 100
Change Not Identified 102
Verification of Design/Field Changes LTA (No PSSR) 103
Documentation Content LTA or Not Kept Up-to-Date 104
Control of Official Documents LTA 105
Customer Interface Services 106
Customer Requirements Not Identified 108
Customer Needs Not Addressed 109
Implementation LTA 110

Procedures 111
Not Used 112
Not Available or Inconvenient to Obtain 113
Procedure Difficult to Use 114

Index of Root Cause Map Nodes by Area (continued)

Use Not Required but Should Be 115
No Procedure for Task 116
Misleading/Confusing 117
Format Confusing or LTA 118
More Than One Action per Step 120
No Checkoff Space Provided but Should Be 121
Inadequate Checklist 122
Graphics LTA 123
Ambiguous or Confusing Instructions/Requirements 124
Data/Computations Wrong/Incomplete 125
Insufficient or Excessive References 126
Identification of Revised Steps LTA 127
Level of Detail LTA 128
Difficult to Identify 129
Wrong/Incomplete 130
Typographical Error 131
Sequence Wrong 132
Facts Wrong/Requirements Not Correct 133
Wrong Revision or Expired Procedure Revision Used 134
Inconsistency Between Requirements 135
Incomplete Situation Not Covered 136
Overlap or Gaps Between Procedures 137

Human Factors Engineering 138
Workplace Layout 140
Controls/Displays LTA 141
Control/Display Integration Arrangement LTA 143
Location of Controls/Displays LTA 144
Conflicting Layout 145
Equipment Location LTA 146
Labeling of Equipment or Locations LTA 147
Work Environment 148
Housekeeping LTA 149
Tools LTA 150
Protective Clothing/Equipment LTA 151
Ambient Conditions LTA 152
Other Environmental Stresses Excessive 154
Workload 155
Excessive Control Action Requirements 156
Unrealistic Monitoring Requirements 157
Knowledge-based Decision Required 158
Excessive Calculation or Data Manipulation Required 159

Index of Root Cause Map Nodes by Area (continued)

Intolerant System 160
Errors Not Detectable 161
Errors Not Correctable 162

Training 163
No Training 164
Decision Not to Train 165
Training Requirements Not Identified 166
Training Records System LTA 167
Training Records Incorrect 168
Training Records Not Up-to-Date 169
Training LTA 170
Job/Task Analysis LTA 171
Program Design/Objectives LTA 172
Lesson Content LTA 174
On-the-job Training LTA 175
Qualification Testing LTA 176
Continuing Training LTA 177
Training Resources LTA 178
Abnormal Events/Emergency Training LTA 179

Immediate Supervision 180
Preparation 181
No Preparation 182
Job Plan LTA 183
Instructions to Workers LTA 184
Walkthrough LTA 185
Scheduling LTA 186
Worker Selection/Assignment LTA 187
Supervision During Work 188
Supervision LTA 189
Improper Performance Not Corrected 190
Teamwork LTA 191

Communications 192
No Communication or Not Timely 194
Method Unavailable or LTA 195
Communication Between Work Groups LTA 196
Communication Between Shifts and Management LTA 197
Communication with Contractors LTA 198
Communication with Customers LTA 199

Misunderstood Communication 200
Standard Terminology Not Used 201
Verification/Repeat-back Not Used 202
Long Message 203
Wrong Instructions 204
Job Turnover LTA 205
Communication Within Shifts LTA 206
Communication Between Shifts LTA 207

Personal Performance 208
Problem Detection LTA 209
Sensory/Perceptual Capabilities LTA 210
Reasoning Capabilities LTA 211
Motor/Physical Capabilities LTA 212
Attitude/Attention LTA 213
Rest/Sleep LTA (Fatigue) 214
Personal/Medication Problems 215

Alphabetized Index of Root Cause Map Nodes

Abnormal Events/Emergency Training LTA 179
Activity Implementation LTA (Predictive Maintenance LTA) 40
Activity Implementation LTA (Preventive Maintenance LTA) 35
Activity Implementation LTA (Proactive Maintenance LTA) 45
Activity Implementation LTA (Routine Equipment Rounds LTA) 54
Administrative/Management Systems 55
Allocation or Resources LTA 27
Ambient Conditions LTA 152
Ambiguous or Confusing Instructions/Requirements 124
Analysis/Design Procedure LTA 24
Attitude/Attention LTA 213
Audits LTA 82

Change Not Identified 102
Communication Between Shifts and Management LTA 197
Communication Between Shifts LTA 207
Communication Between Work Groups LTA 196
Communication of SPACs LTA 69
Communication with Contractors LTA 198
Communication with Customers LTA 199
Communication Within Shifts LTA 206
Communications 192
Conflicting Layout 145
Confusing, Contradictory, or Incomplete 61
Continuing Training LTA 177
Contractor Selection LTA 99
Control/Display Integration/Arrangement LTA ... 143
Control of Changes to Procurement Specifications LTA 96
Control of Official Documents LTA 105
Controls/Displays LTA 141
Corrective Action LTA 83
Corrective Actions Not Yet Implemented 84
Company Employee 9
Contractor Employee 10
Customer Interface Services 106
Customer Needs Not Addressed 109
Customer Requirements Not Identified 108

Data/Computations Wrong/Incomplete 125
Decision Not to Train 165
Design Input LTA 16
Design Input/Output 15
Design Output LTA 17
Detection LTA 37
Difficult to Identify 129
Document and Configuration Control 100
Documentation Content LTA or Not Kept Up-to-Date 104

Employee Screening/Hiring LTA 66
Enforcement LTA 71
Equipment Design Problem 5
Equipment Design Records LTA 19
Equipment Difficulty 2
Equipment Location LTA 146
Equipment Operating/Maintenance History LTA 20
Equipment Records 18
Equipment Reliability Program Design LTA 21
Equipment Reliability Program Implementation LTA 28
Equipment Reliability Program Problem LTA 6
Errors Not Correctable 162
Errors Not Detectable 161
Event Specification LTA 42
Excessive Calculations or Data Manipulation Requred 159
Excessive Control Action Requirements 156

Facts Wrong/Requirements Not Correct 133
Failure Finding Maintenance LTA 46
Format Confusing or LTA 118
Frequency LTA (Failure Finding Maintenance LTA) 47
Frequency LTA (Preventive Maintenance LTA) 33
Frequency LTA (Routine Equipment Rounds LTA) 52

Alphabetized Index of Root Cause Map Nodes (continued)

Graphics LTA 123

Handling LTA 87
Housekeeping LTA 149
Human Factors Engineering 138

Identification of Revised Steps LTA 127
Immediate Supervision 180
Implementation LTA 110
Improper Performance Not Corrected 190
Inadequate Checklist 122
Inappropriate Type of Maintenance Assigned 25
Incomplete/Situation Not Covered 136
Inconsistency Between Requirements 135
Installation/Fabrication 7
Instructions to Workers LTA 184
Insufficient or Excessive References 126
Intolerant System 160

Job Plan LTA 183
Job/Task Analysis LTA 171
Job Turnover LTA 205

Knowledge-based Decision Required 158

Labeling of Equipment or Locations LTA 147
Lesson Content LTA 174
Level of Detail LTA 128
Location of Controls/Displays LTA 144
Long Message 203

Material Acceptance Requirements LTA 97
Material Inspections LTA 98
Method Unavailable or LTA 195
Misleading or Confusing 117
Misunderstood Communication 200
Monitoring LTA (Predictive Maintenance LTA) 38
Monitoring LTA (Proactive Maintenance LTA) 43
More Than One Action per Step 120
Motor/Physical Capabilities LTA 212

Natural Phenomena 11
No Checkoff Space Provided but Should Be 121
No Communication or Not Timely 194
No Preparation 182
No Procedure for Task 116
No Program 22
No SPACs 59
No Training 164
Not Available or Inconvenient to Obtain 113
Not Strict Enough 60
Not Used 112

On-the-job Training LTA 175
Other 14
Other Difficulty 4
Other Environmental Stresses Excessive 154
Overlap or Gaps Between Procedures 137

Packaging/Shipping LTA 89
Personal/Medication Problems 215
Personal Performance 208
Personnel Difficulty 3
Planning, Scheduling, or Tracking of Work Activities LTA 64
Predictive Maintenance LTA 36
Preparation 181
Preventive Maintenance LTA 32
Proactive Maintenance LTA 41
Problem Analysis LTA 81
Problem Detection LTA 209
Problem Identification/Control 78
Problem Reporting LTA 80
Procedure Difficult to Use 114
Procedures 111
Procurement Control 93
Product Acceptance Criteria LTA 91
Product Inspections LTA 92
Program Design/Objectives LTA 172
Program LTA 23
Protective Clothing/Equipment LTA 151
Purchasing Specifications LTA 95

Alphabetized Index of Root Cause Map Nodes (continued)

Qualification Testing LTA 176
Reasoning Capabilities LTA 211
Recently Changed 70
Recommendations Not Yet Implemented 75
Repair Implementation LTA (Corrective Maintenance LTA) 31
Repair Implementation LTA (Failure Finding Maintenance LTA) 50
Responsibility for Item/Activity Not Adequately Defined 63
Rest/Sleep LTA (Fatigue) 214
Review LTA or Not Performed 74
Review Procedure LTA 77
Rewards/Incentives LTA 65
Risk Acceptance Criteria LTA (Program LTA) 26
Risk Acceptance Criteria LTA (Safety/Hazard/Risk Review) 76
Routine Equipment Rounds LTA 51

Sabotage/Horseplay 12
Safety/Hazard Review 72
Scheduling LTA 186
Scope LTA (Failure Finding Maintenance LTA) ... 48
Scope LTA (Preventive Maintenance LTA) 34
Scope LTA (Proactive Maintenance LTA) 44
Scope LTA (Routine Equipment Rounds LTA) 53
Sensory/Perceptual Capabilities LTA 210
Sequence Wrong 132
Standard Terminology Not Used 201
Standards, Policies, or Administrative Controls (SPACs) Not Used 67
Standards, Policies, or Administrative Controls (SPACs) LTA 57
Storage LTA 88
Supervision During Work 188
Supervision LTA 189

Teamwork LTA 191
Technical Error 62
Tools LTA 150
Training 163
Training LTA 170

Training Records Incorrect 168
Training Records Not Up-to-Date 169
Training Records System LTA 167
Training Requirements Not Identified 166
Training Resources LTA 178
Troubleshooting/Corrective Action LTA (Corrective Maintenance LTA) 30
Troubleshooting/Corrective Action LTA (Failure Finding Maintenance LTA) 49
Troubleshooting/Corrective Action LTA (Predictive Maintenance LTA) 39
Typographical Error 131

Unauthorized Material Substitution 90
Unrealistic Monitoring Requirements 157
Use Not Required but Should Be 115

Verification of Design/Field Changes LTA (No PSSR) 103
Verification/Repeat-back Not Used 202

Walkthrough LTA 185
Work Environment 148
Worker Selection/Assignment LTA 187
Workload 155
Workplace Layout 140
Wrong/Incomplete 130
Wrong Instructions 204
Wrong Revision or Expired Procedure Revision Used 134

Typical Issues

Start here with each causal factor. See the following pages for attributes of each of the next three nodes, which will help you choose the correct one.

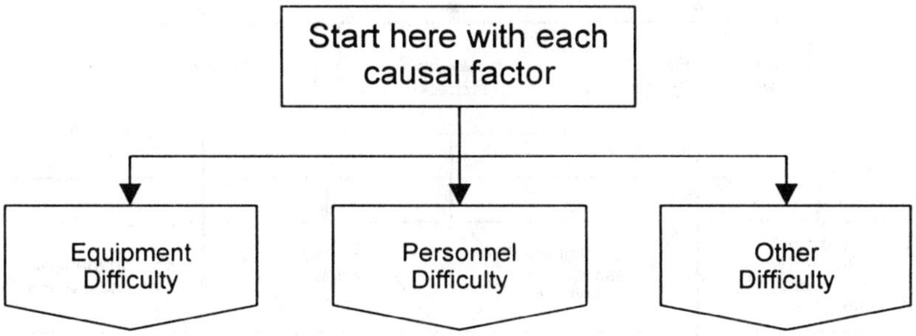

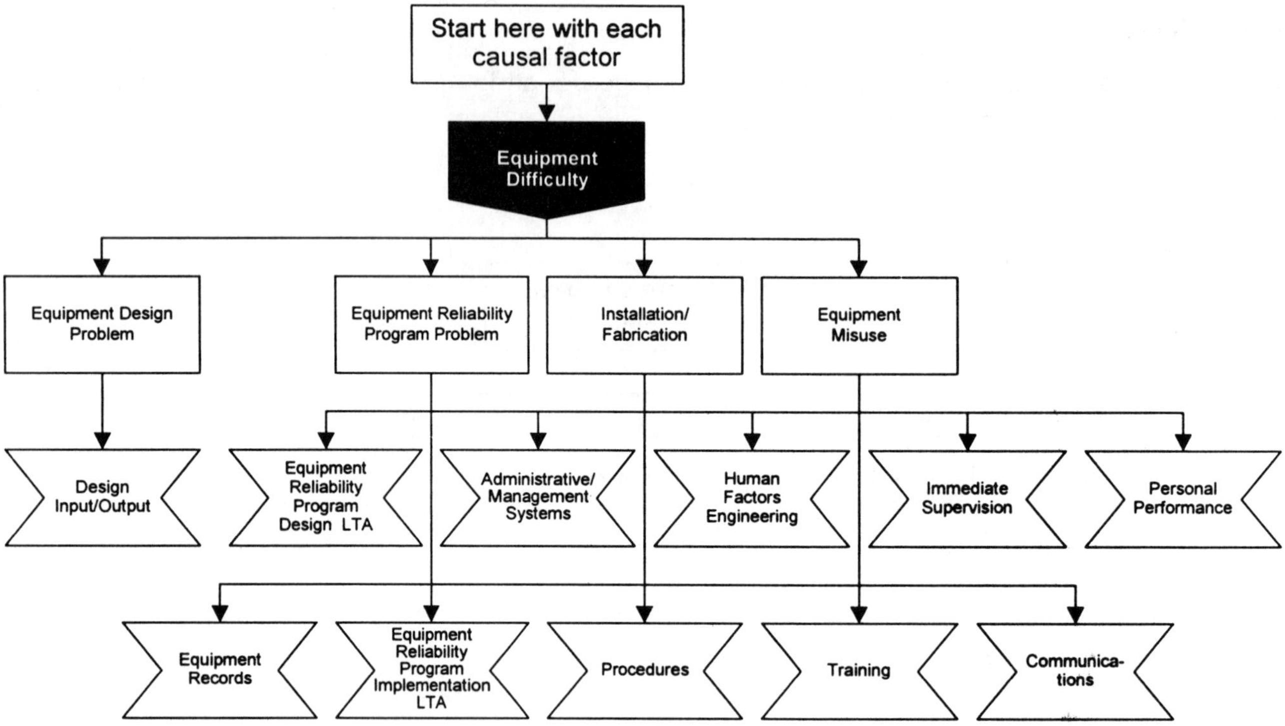

Typical Issues

These include problems with equipment design, fabrication, installation, maintenance, and misuse. Problems with the equipment reliability program are also identified/categorized under this node.

Typical Recommendation

See lower level nodes.

Example

A spill to the environment occurred because a valve failed. The valve failed because it was not designed for the environment in which it operated.

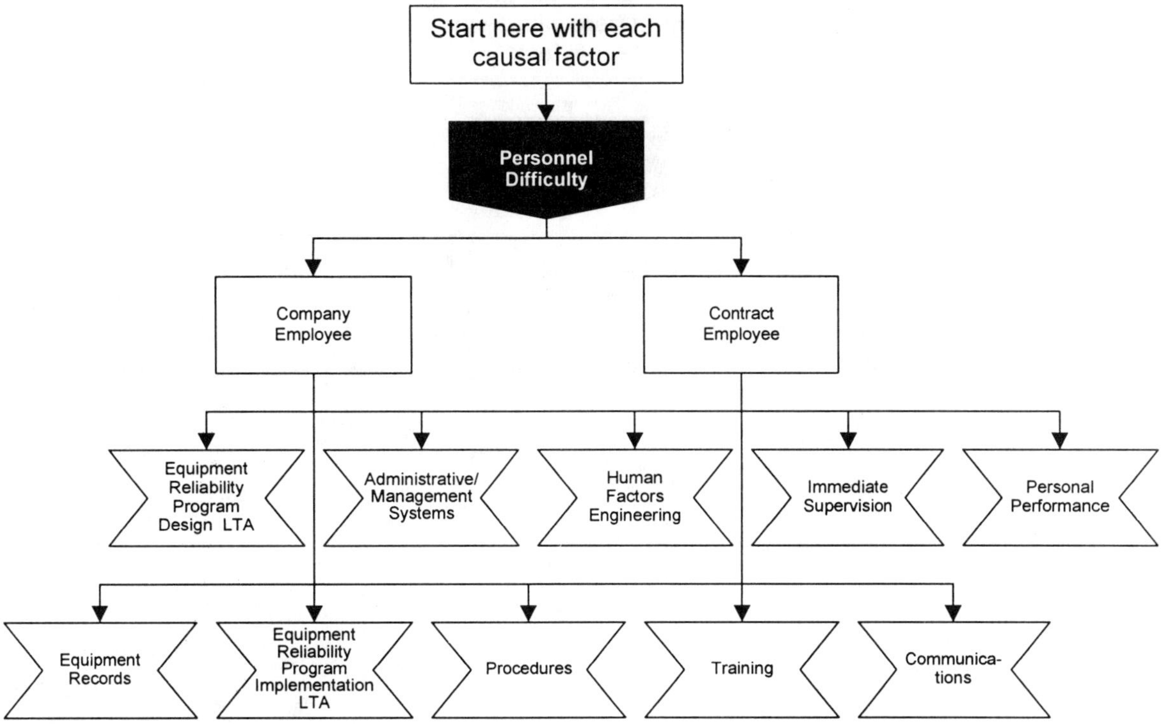

Typical Issues

Problems related to the "running" of the plant/process are identified/categorized under this node. This includes the administrative/management systems to control the process (e.g., standards, policies, procedures), training of personnel, communications, and communications among individuals and groups.

Typical Recommendation

See lower level nodes.

Examples

A tank overflowed, resulting in a spill to the environment. The operator filling the tank was using the wrong revision of the procedure, which had an incorrect calibration chart for the tank.

A mechanic doing maintenance in a confined space was not allowed to take a written procedure with him. As a result, he had to review the procedure and commit it to memory. During performance of the task, he omitted an important step. This resulted in the failure of a key piece of equipment.

An operator made a mistake performing a calculation. The data used in the calculation came from multiple steps in the procedure. He made a mistake in transferring one of the data points from an earlier step in the procedure to the step at which the calculation was performed.

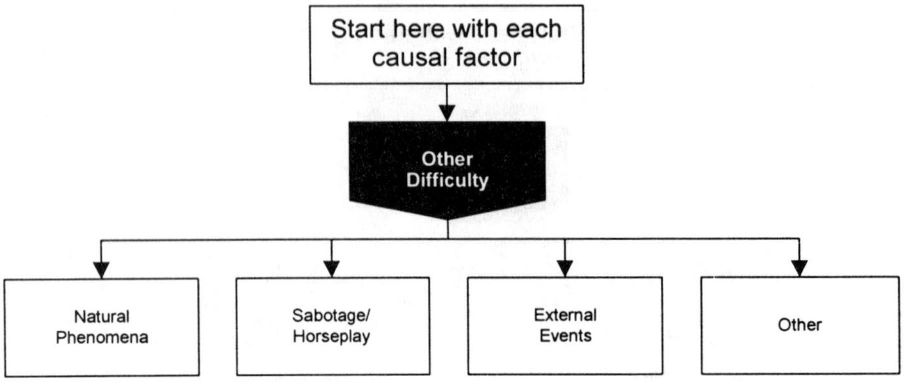

Typical Issues

These include problems related to natural phenomena, sabotage, external events, and events that cannot be categorized elsewhere.

Typical Recommendation

See lower level nodes.

Examples

Inventory in the warehouse was damaged when the warehouse was flooded following a heavy rain.

A mechanic intentionally damaged a lathe. He was disgruntled about being placed into a new assignment.

A release of chlorine from an adjacent facility affected the operators in your facility.

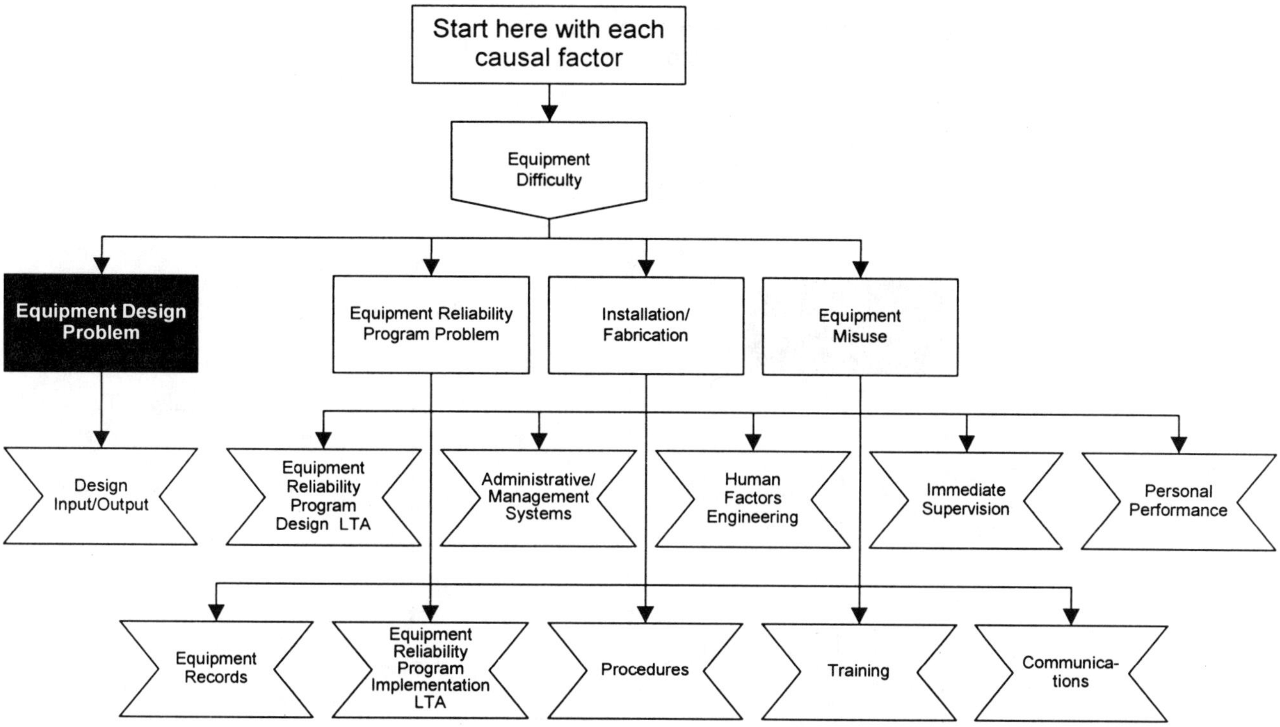

Typical Issues

These include problems related to the design process, problems related to the design and capabilities of the equipment, and problems related to the specification of parts and materials.

Typical Recommendation

See lower level nodes.

Examples

A valve failed because the designer used obsolete materials requirements.

A process upset occurred because one of the flow streams was out of specification. The design input did not indicate all the possible flow rates for the process. The pump was incorrectly sized for the necessary flow requirements.

A line ruptured because a gasket failed. The gasket was constructed of the wrong material because the design did not consider all the possible chemicals that would be in the line during different operating conditions. A chemical that was not considered caused the gasket to fail.

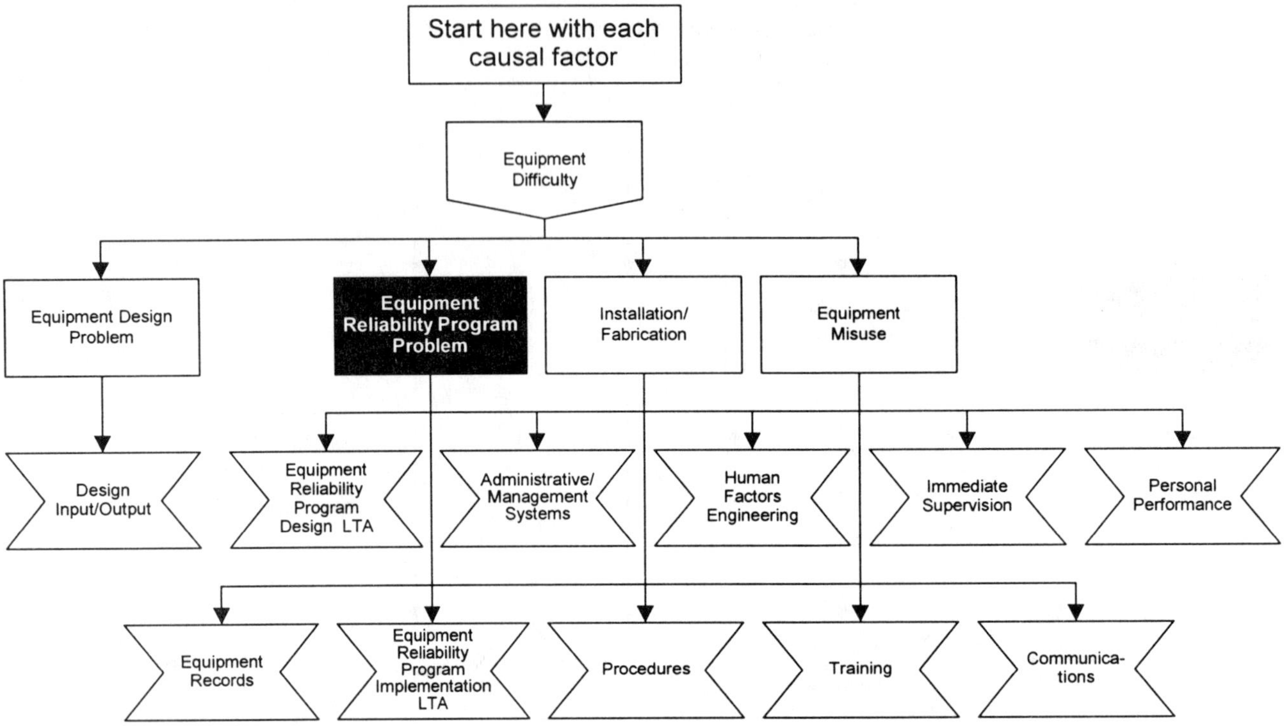

Typical Issues

These include problems related to the design and implementation of the maintenance program. Was the wrong type of maintenance specified for the equipment? Are there problems with the analysis process used to determine the appropriate maintenance requirements? Are there problems related to performing the maintenance activities? Are monitoring activities implemented to detect deteriorating equipment? Does the repair activity cover the required scope?

Typical Recommendations

- Improve equipment operational and maintenance records to enable selection of the proper type of maintenance
- Assign additional resources to equipment with a demonstrated history of problems
- Reduce maintenance on equipment that has no significant impact on production or safety and that can be easily repaired or replaced
- Provide maintenance procedures and training appropriate to the experience level of personnel

Examples

During the past year, the failure rate for the feed pumps has doubled. Maintenance records are inadequate to determine why any of the failures occurred. Work records just say "pump repaired."

A number of pump bearings have failed recently. Predictive maintenance was selected as the appropriate type of maintenance for the pump bearings. However, there is no requirement for monitoring of the pump bearings. As a result, the predictive maintenance activity was never implemented.

Preventive maintenance (a calibration) was being performed on a product scale every 3 months. However, operators requested additional calibrations about once per month as they noticed the scale drifting. The frequency of the calibration was changed to once per month.

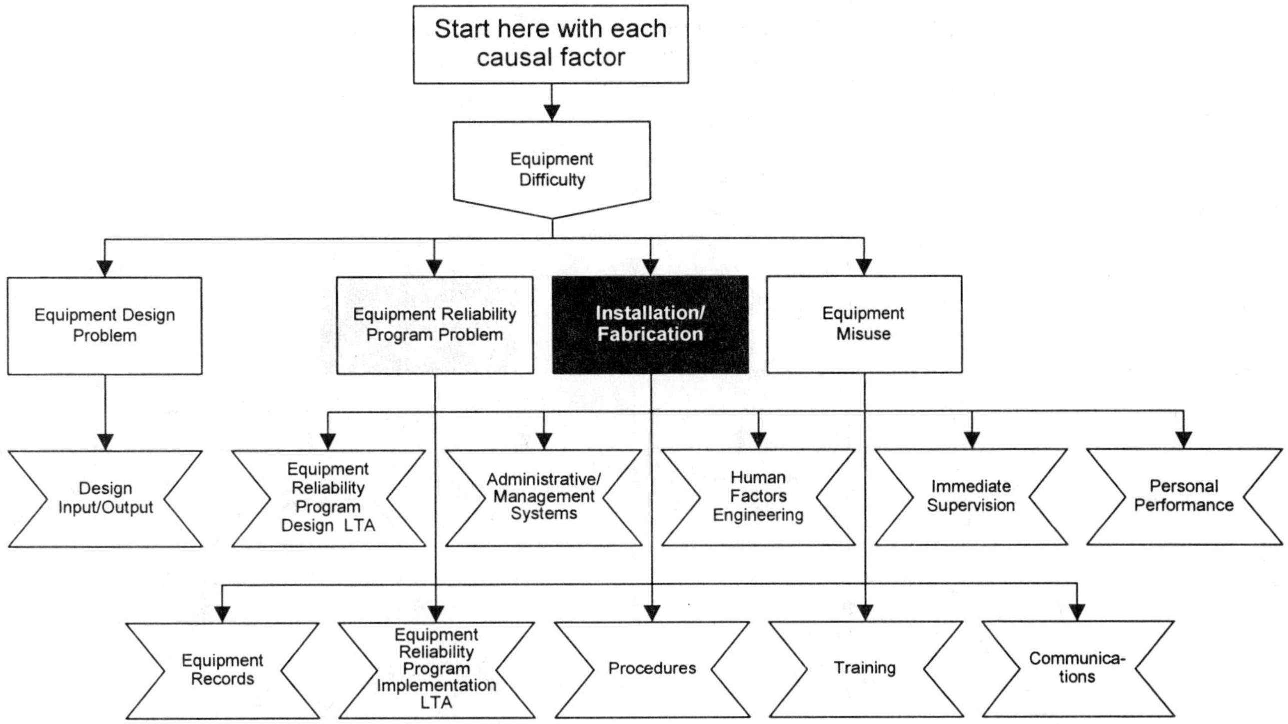

Typical Issues

These include problems with the fabrication and installation of equipment. Was the equipment fabricated to the design specifications? Was there a problem with the installation? Was a field modification performed that altered the performance of the equipment?

Typical Recommendations

- Develop fabrication procedures to help ensure that the fabricated equipment meets design specifications
- Involve design personnel in field reviews of the fabrication and installation of equipment
- Allow field fabrication and installation personnel to have access to design personnel to resolve problems encountered in the fabrication and installation process

Examples

A pipe needed to be rerouted during installation to go around existing equipment, but this was not on the layout drawing. The reroute created a low point in the line that allowed contaminants to accumulate. Later, the pipe failed in this low section.

Field personnel could not determine from the installation package how to connect the power to a new drill press. They decided to connect it the same way the others in the facility were connected even though this drill press had a different manufacturer than the rest. As a result of the incorrect connection, the drill press control system was damaged.

A walkway collapsed because a field modification of the suspension system resulted in a weakened support system. The walkway collapsed when it was full of people.

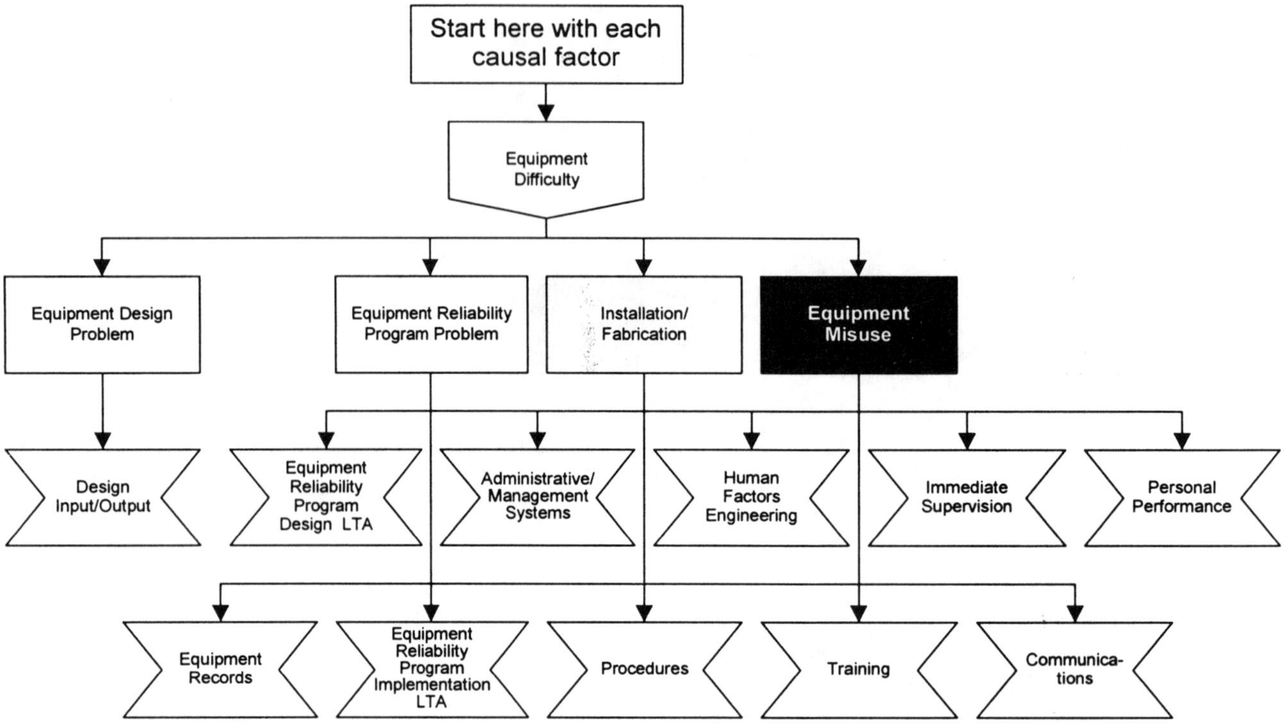

Typical Issues

Was equipment used for an activity other than it was intended? Was the equipment used beyond its capacity?

Typical Recommendations

- Ensure that proper equipment is available for personnel to use
- Ensure that personnel are aware of the proper use of equipment

Examples

To save money, a drill press was purchased to mix chemicals in a lab. The slowest speed on the drill press was still too fast for proper mixing of materials. As a result, technicians were routinely splattered with chemicals while using the drill press.

A technician in the field ran out of a water-based cleaner. Someone nearby was using gas to run a lawn mower. Instead of going back to the truck to get the water-based cleaner, the technician put some gas on a rag and used it. The operator's hair was burned when the rag contacted a hot bearing, starting a fire.

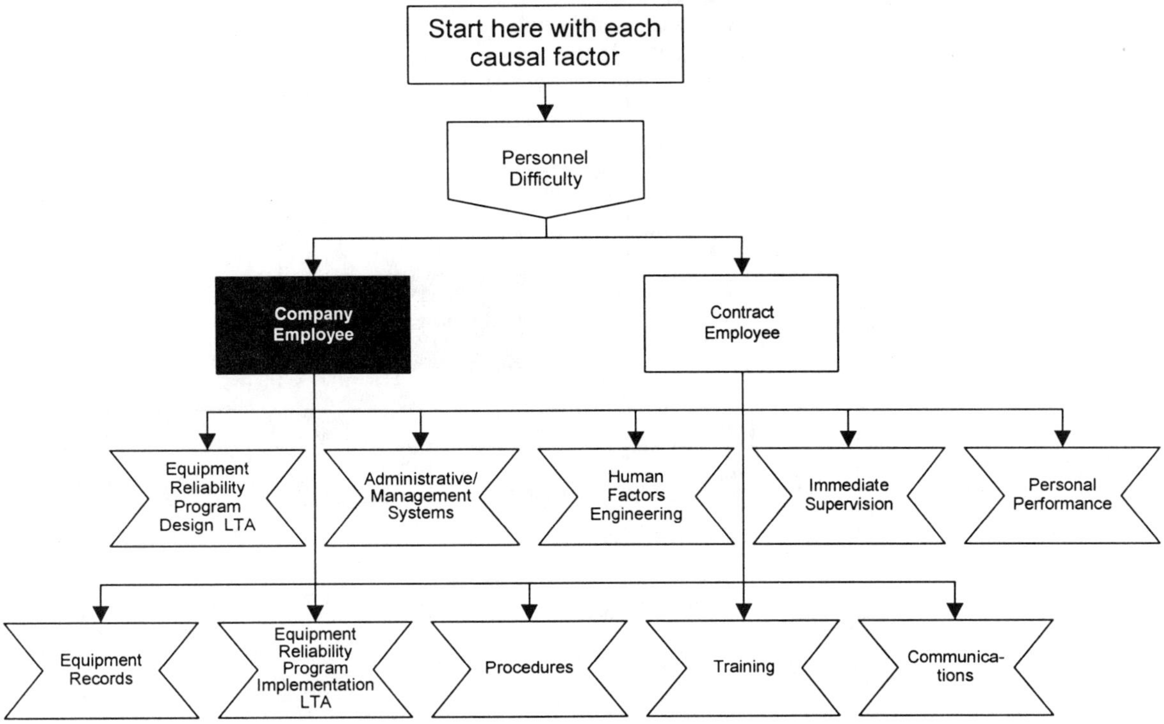

Typical Issues

Was a company employee involved? Are the employees involved covered by the normal company training programs? Are they supervised directly by company employees?

Note: *Distinguishing between company and contract employees can be important because of the different management systems that control the work performed by these two groups.*

Typical Recommendation

See lower level nodes.

Examples

A company employee took a sample from product tank C instead of product tank B. The tanks are arranged from left to right: A C B.

A company employee made a mistake using a scale to weigh a pallet of material. It was the first time the operator had used the scale. He was told how to use the scale as part of training, but had never actually used it himself.

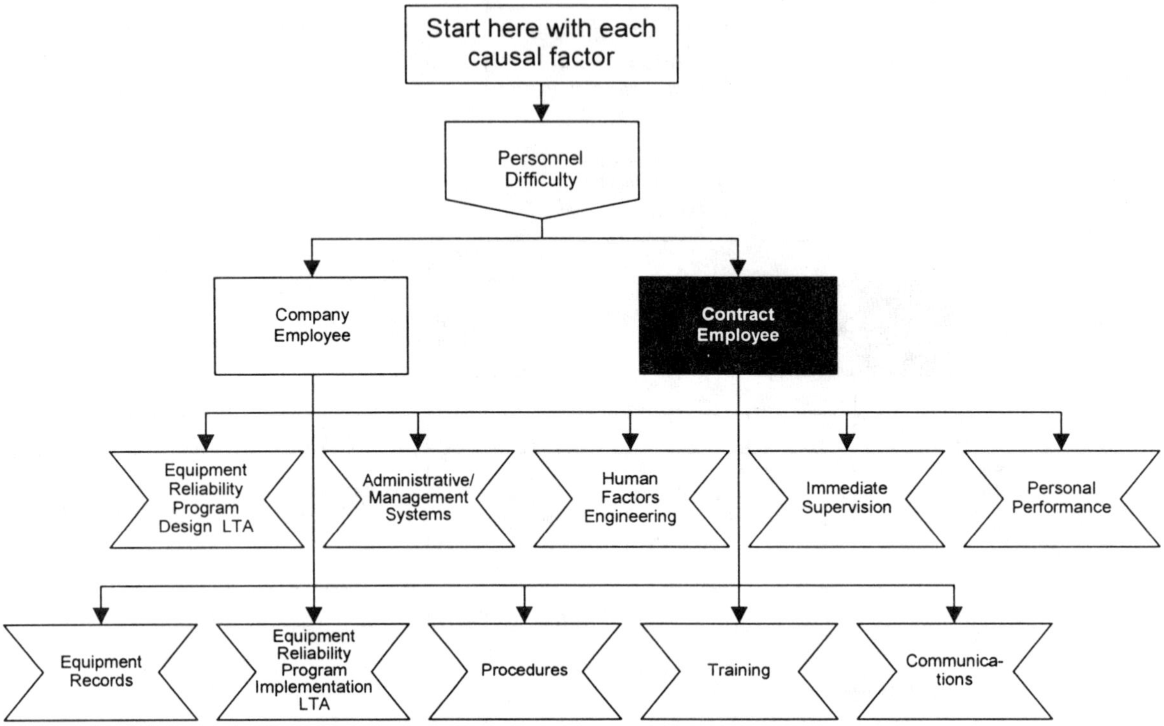

Typical Issues

Was a contract employee involved? Was the person involved a transient worker in your facility? Was the person involved covered by the contract employee training program? Is the person involved directly supervised by someone who does not work for your company? Does this person have to meet different requirements than a "regular" employee?

Note: *Distinguishing between company and contract employees can be important because of the different management systems that control the work performed by these two groups.*

Typical Recommendations

- Ensure that contract employees have sufficient guidance to perform their activities
- Ensure that work documents used by contract employees have sufficient detail to allow individuals inexperienced with your operations and work methods to adequately perform the job

Examples

A worker for the local vending company entered the facility to refill the vending machines. The individual was not aware of the requirement to wear a hard hat and safety goggles in the aisle way that led to the lunchroom.

A contract mechanic installed the wrong type of gasket in a line during a scheduled maintenance activity. As a result, the line failed when the process was restarted. The procedure did not specify the proper material to be used. The in-house mechanics all knew the proper material, and, therefore, it had never been a problem even though it was not specifically covered in the procedure.

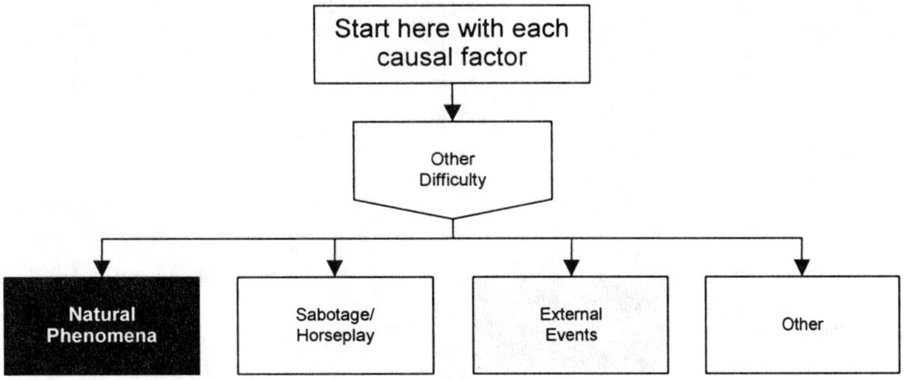

Typical Issues

Problems that result from tornadoes, hurricanes, earthquakes, lightning, floods, or other natural phenomena are identified under this node.

Typical Recommendations

- Ensure that natural phenomena are considered in the design process
- Ensure that natural phenomena are considered in the development of procedures and training
- Ensure that risk acceptance criteria are properly set for natural phenomena events

Examples

A process upset occurred in the facility because power was lost as a result of lightning striking a transformer.

The plant site was flooded when the river overtopped the levee designed for a 100-year flood.

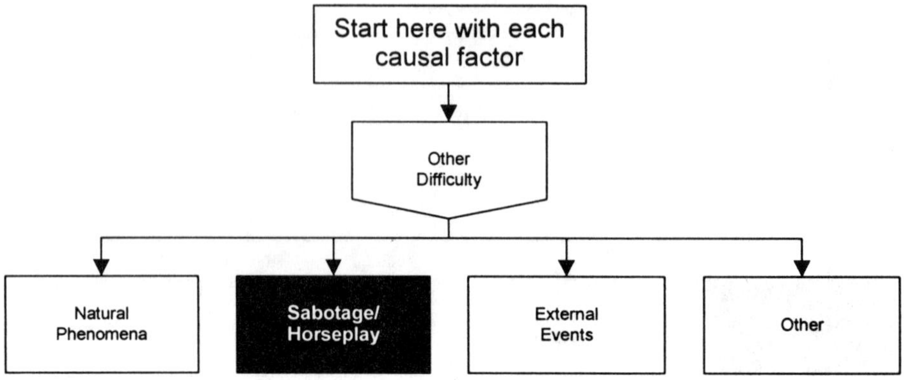

Typical Issues

Malicious acts that cause or contribute to an incident are identified under this node. Malicious lack of action that contributed to a problem is also identified under Sabotage. Was the event the result of horseplay or other nonwork-related activities?

Note: *Dual coding under* Personal Performance *or* Administrative/Management Systems *may be appropriate. The problem that led the employee to commit sabotage/horseplay should be addressed.*

Typical Recommendations

- Ensure that security plans and equipment are adequate
- Ensure that inappropriate behavior, such as horseplay, is addressed and corrected by management and supervision

Examples

A mechanic intentionally damaged a piece of equipment. He was disgruntled about being placed in a new assignment.

As a practical joke, operators sent a junior operator to check out the *electrical zerts* (there are no such things) on the generator. As a result of trying to find the *electrical zerts*, the junior operator accidentally shut down the generator.

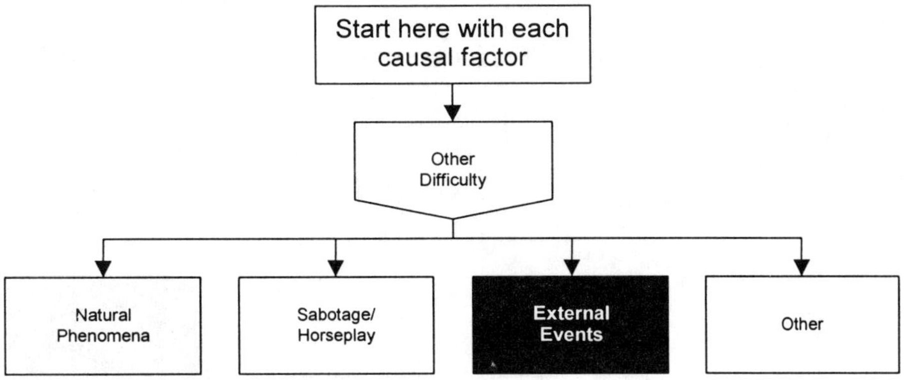

Typical Issues

Was the event a result of problems at adjacent facilities? Was it the result of activities outside the facility that are not under your control?

Note: *Coding under* Administrative/Management Systems; Safety/Hazard/Risk Review *may also be appropriate.*

Typical Recommendations

- Coordinate emergency response and planning with nearby facilities
- Develop contingency actions for external events

Examples

A chlorine tanker accident on a nearby railroad spur requires the evacuation of a portion of your facility.

A nearby accident on the expressway prevents shipments from leaving your facility for an 8-hour period. As a result, some deliveries are not made on time.

A key supplier's warehouse was struck by a tornado. As a result, the warehouse was unable to supply your facility with raw materials for a period of 2 weeks.

The local utility's power plant shut down, resulting in a 5-minute power outage to your facility. It took 2 hours to restart the plant and stabilize the process.

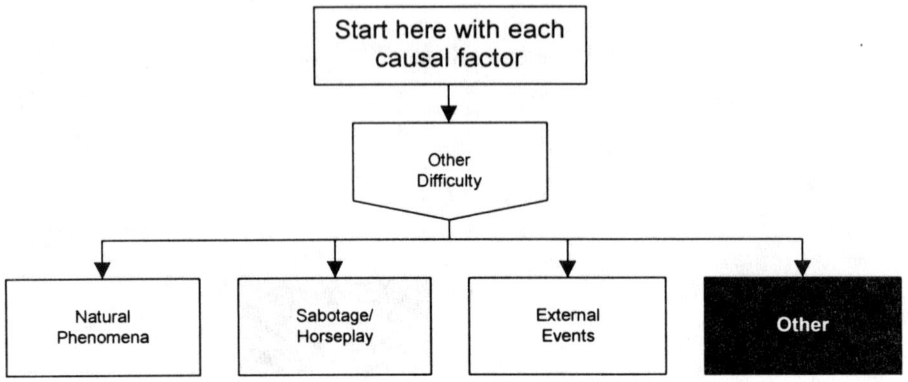

Typical Issues

These include issues that cannot be coded elsewhere on the map (e.g., problems that cannot be coded because of insufficient information).

Typical Recommendations

- Analyze the causal factors that are coded under this node. Determine if additional nodes should be added to the map to categorize these issues
- Determine methods for gathering additional information for this type of event when it recurs

Examples

A customer complained that the materials sent to him were out of specification. However, when the lab sample was tested, it was acceptable. When the customer retested the material, his test also indicated the material was acceptable. Product manufactured from this batch was also acceptable.

A spurious shutdown of a computer in the order receiving department caused a delay in handling a customer's request. The problem could not be recreated. It could not be determined whether it was equipment failure or human error that led to the shutdown.

A rabid fox bit a worker who was checking some equipment in a remote location.

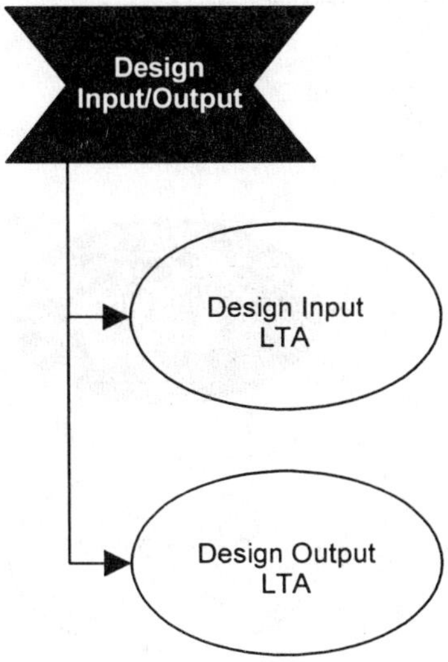

Typical Issues

Were all the appropriate design inputs considered during the design phase? Was the design output, such as drawings and specifications, complete? Was the design input and output consistent and complete? Did the design review process fail to detect errors? Was the design review independent?

Typical Recommendations

- Conduct a feasibility review prior to beginning design to ensure that the criteria can be met and that no conflicting criteria exist
- Develop a pre-construction planning and review process to help ensure that all the specifications are in agreement

Examples

A valve failed because equipment conditions during operation, such as corrosivity, were not considered during design.

A pump failed to deliver enough cooling water in an emergency because emergency requirements were not considered in the design.

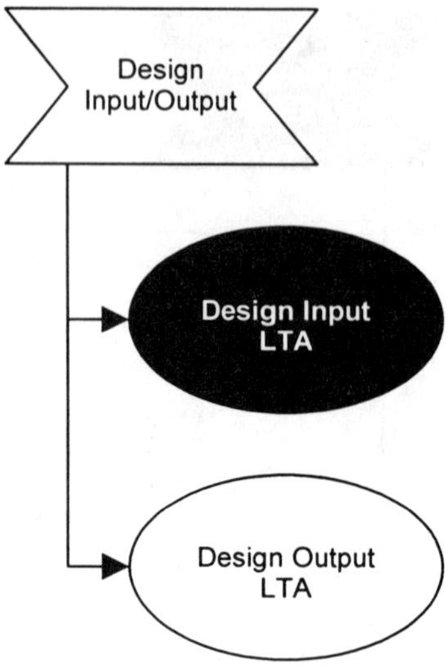

Typical Issues

Were all the appropriate design inputs considered during the design phase? Were the design criteria so stringent that they could not be met? Were some criteria conflicting? Were requirements out-of-date? Were the wrong standards or bases used? Were the necessary codes and standards available to the designer?

Typical Recommendations

- Conduct a feasibility study prior to beginning design to ensure that the criteria can be met and that no conflicting criteria exist
- Develop an independent review process to help ensure that appropriate standards are used in the design
- Develop a tracking system to help ensure that design problems and conflicts are resolved prior to startup
- Develop a tracking system to help ensure that current design criteria are used
- Develop comprehensive system design requirements

Examples

A valve failed because the designer used obsolete materials requirements.

A process upset occurred because one of the flow streams was out of specification. The design input did not indicate all the possible flow rates for the process. The pump was incorrectly sized for the necessary flow requirements.

An engineer did not account for all types of vehicles that would be required to enter the plant in the design of the new guard house and gate. As a result, some of the outside responder's fire trucks can no longer enter the plant because they are wider than the new entrance.

A flow controller could not adequately control flow during an infrequent operation. The flow requirements for normal, emergency, and infrequent operation covered too wide a range for a controller to operate properly under all of the conditions.

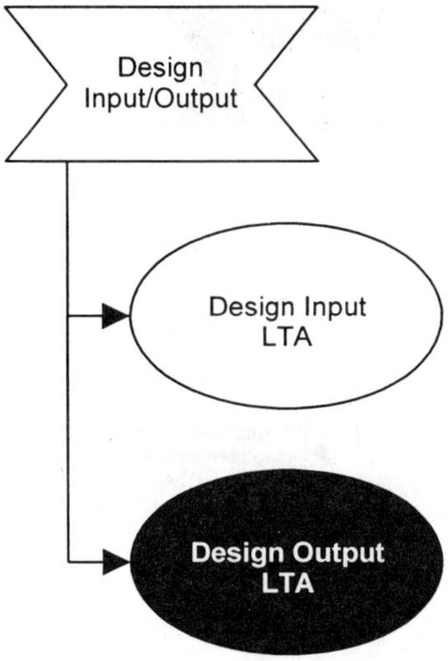

Typical Issues

Was the design output, such as drawings and specifications, complete? Were all operating conditions (normal, startup, shutdown, emergency) considered in the design? Were the design documents difficult to read or interpret? Did the final design output include all changes? Were there differences between different output documents? Did the design output address all requirements specified in the design input?

Typical Recommendations

- Include satisfaction of design input criteria as a specific review team item during design reviews
- Include experienced operations and maintenance personnel in design reviews to help ensure that all possible operating conditions are considered in the design
- Include designers in construction and pre-startup reviews to help ensure that design information is properly interpreted
- Conduct an independent technical review of the final design to help ensure consistency among various design documents

Examples

A valve failed because the material specifications were incorrect. The specifications did not agree with the design criteria. The criteria stated that the valve must operate in a corrosive environment, but the specifications did not indicate this condition. Therefore, the valve was constructed of improper materials.

A line ruptured because a gasket failed. The gasket was constructed of the wrong material because the design did not consider all the possible chemicals that would be in the line during different operating conditions. One that was not considered caused the gasket to fail.

A pump did not provide the necessary cooling water during an emergency. The pump was sized incorrectly because the final design specifications did not include changes identified in the safety analysis.

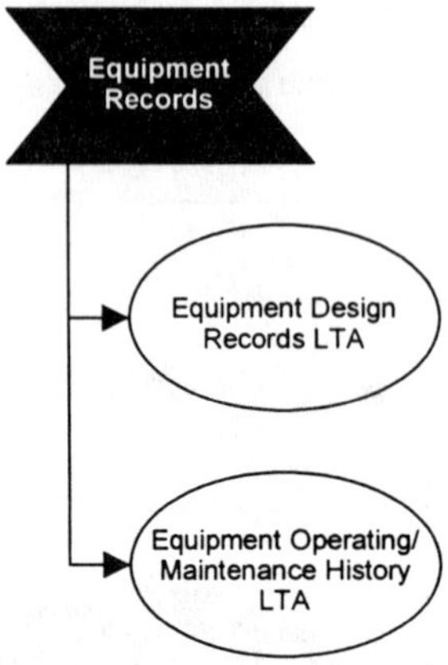

Typical Issues

Does an equipment records program exist? Is it adequate and up to date? Does it contain the correct information? Does it contain all the information necessary to ensure equipment reliability?

Typical Recommendations

- Develop a system for tracking equipment histories
- Collect information from other sources (e.g., vendors) to help complete existing equipment histories

Examples

A tank overflowed because of faulty liquid level instrumentation. The records indicated that a calibration was called for and performed 3 months prior, but did not indicate how much adjustment was made during calibration. A large adjustment might have indicated pending failure.

A pressure vessel was not properly tested after a modification. The design information for the salvaged vessel had been lost.

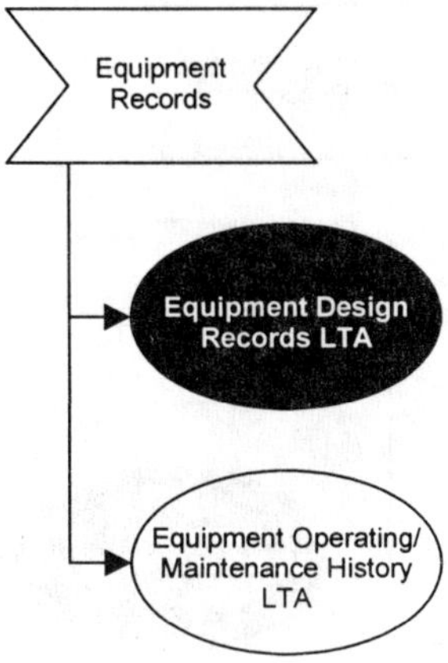

Typical Issue

Have problems with design records caused problems with the operation, maintenance, or modification of equipment?

Typical Recommendation

- Ensure that design information is retained on equipment and accessible to personnel responsible for operation, maintenance, and modification of the equipment

Examples

As part of a capacity upgrade, engineers attempted to determine the design throughput of a blender. No equipment records could be located to determine the design capacity of the equipment.

Maintenance procedures were being developed for a new freezer. Lack of design information required extensive field verification of equipment configuration to develop the procedure.

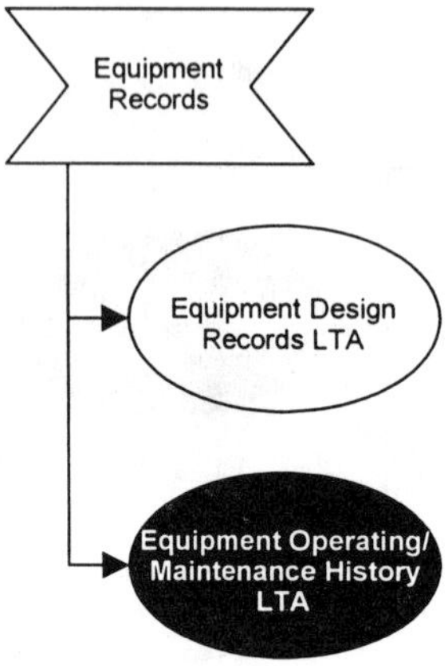

Typical Issues

Was the history for the equipment that malfunctioned complete? Did the history contain information about similar equipment? Would knowledge of the history of the equipment involved in the event and similar equipment have prevented the incident or lessened its severity?

Typical Recommendations

- Collect available information from other sources (e.g., vendors) to help complete existing equipment histories
- Improve the system for tracking equipment histories to help ensure that all pertinent information is retained
- Assign responsibility for maintaining and analyzing equipment repair and maintenance records
- Periodically audit the equipment history files to help ensure that the records system is being followed
- Assess adequacy of operator rounds and the information collected on rounds
- Assess adequacy of maintenance tasks that collect information on the status of equipment
- Ensure that information collected on rounds is analyzed to determine if problems exist with equipment

Examples

A tank overflowed because of faulty liquid level instrumentation during a nonroutine mode of operation that failed the level device. Previous problems had occurred with the instrumentation under these conditions. This was not known by current personnel because no equipment history was available.

A flow meter in a product line failed, resulting in the wrong amount of material being sent to a customer. Records indicated that calibration of the flow sensor had been performed three times in the last month, but did not indicate how much adjustment was made during calibration. A large adjustment, or larger adjustments each time, might have indicated a pending failure.

Operators routinely performed rounds twice each shift. However, there were no guidelines provided on what to look for or what data to document. Following a number of pump failures, the equipment logs were reviewed to determine what was causing the failures. Although the operators had looked at the pumps each shift, they had not collected any operating history on them.

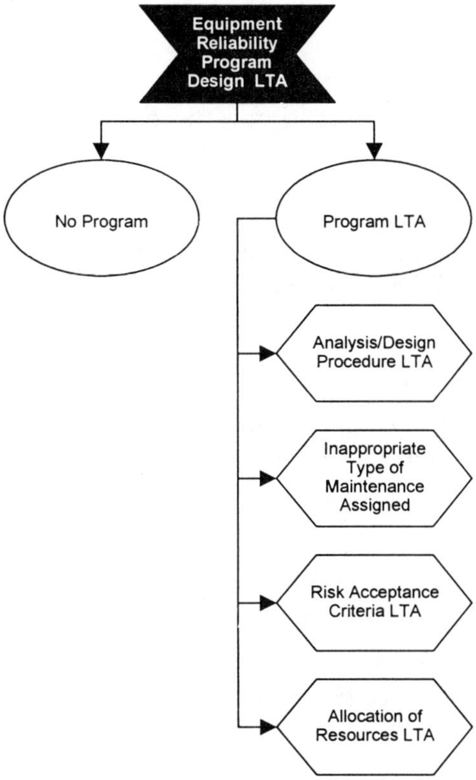

Typical Issues

These include problems related to the design and implementation of the maintenance program. Was the wrong type of maintenance specified for the equipment? Are there problems with the analysis process used to determine the appropriate maintenance requirements?

Typical Recommendations

- Improve equipment operational and maintenance records to enable selection of the proper type of maintenance
- Assign additional resources to equipment with a demonstrated history of problems
- Reduce maintenance on equipment that has no significant impact on production and that can be easily repaired or replaced

Examples

Maintenance activities had been specified for the running components of a wood chipping machine (i.e., bearings, blades) but no maintenance activities had been specified for the safety interlocks associated with the machine. The analysis procedure did not require safety interlocks to be addressed. As a result, an operator's arm was amputated when it was caught in the chipper and the auto stop feature failed.

A number of pump bearings have failed recently. Predictive maintenance was selected as the appropriate type of maintenance for the pump bearings. However, there is no requirement for monitoring of the pump bearings.

Corrective maintenance was assigned to an auger that provided raw materials to a food process. This selection was based on a very low expected failure rate and a quick repair time. Actual experience indicates the failures took much longer to repair than the analysis team estimated. As a result, the risk associated with the failures was much higher than the team thought.

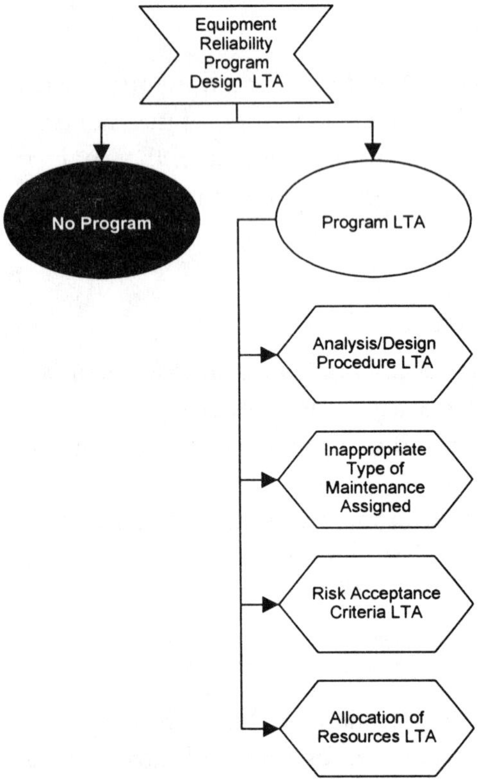

Typical Issues

Does an equipment reliability program exist for this piece of equipment? Have the maintenance needs for this piece of equipment been analyzed?

Note: *If the maintenance needs were analyzed and it was determined that no maintenance was appropriate, code this under* Program LTA (Equipment Reliability Program Design LTA).

Typical Recommendation

- Determine the appropriate level of maintenance for all equipment in the facility that is important to safety or reliability

Example

Hydraulic hoses on the forklifts in the facility were failing once every 2 months. A review of the maintenance program records indicated that proper maintenance for these hoses had never been determined.

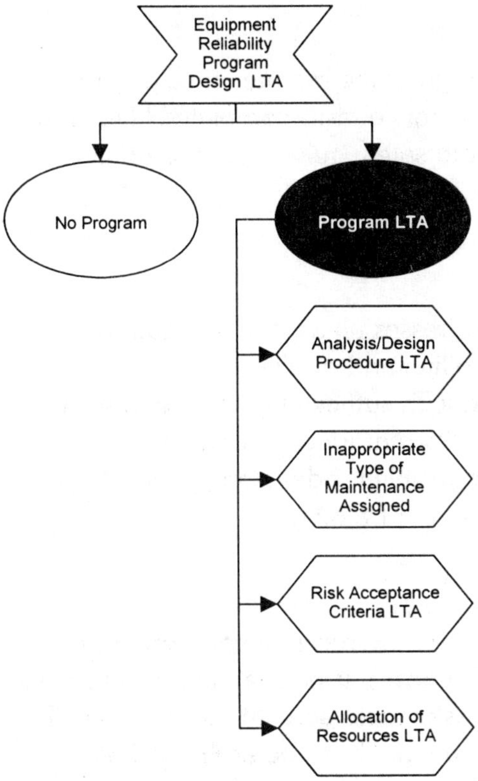

Typical Issues

These include problems related to the design and implementation of the maintenance program. Was the wrong type of maintenance specified for the equipment? Are there problems with the analysis process that is used to determine the appropriate maintenance requirements?

Typical Recommendations

- Ensure that the proper level of risk acceptance is used in determining the level and type of maintenance to perform on equipment
- Ensure that the analysis process addresses all aspects of equipment operation important to safety and reliability
- Improve equipment operational and maintenance records to enable the selection of the proper type of maintenance
- Assign additional resources to equipment with a demonstrated history of problems
- Reduce maintenance on equipment that has no significant impact on production or safety and that can be easily repaired or replaced

Examples

Maintenance activities had been specified for the running components of a wood chipping machine (i.e., bearings, blades) but no maintenance activities had been specified for the safety interlocks associated with the machine. The analysis procedure did not require safety interlocks to be addressed. As a result, an operator's arm was amputated when it was caught in the chipper and the auto stop feature failed.

A number of pump bearings have failed recently. Predictive maintenance was selected as the appropriate type of maintenance for the pump bearings. However, there is no requirement for monitoring of the pump bearings.

Corrective maintenance was assigned to an auger that provided raw materials to a food process. This selection was based on a very low expected failure rate and a quick repair time. Actual experience indicates the failures took much longer to repair than the analysis team estimated. As a result, the risk associated with the failures was much higher than the team thought.

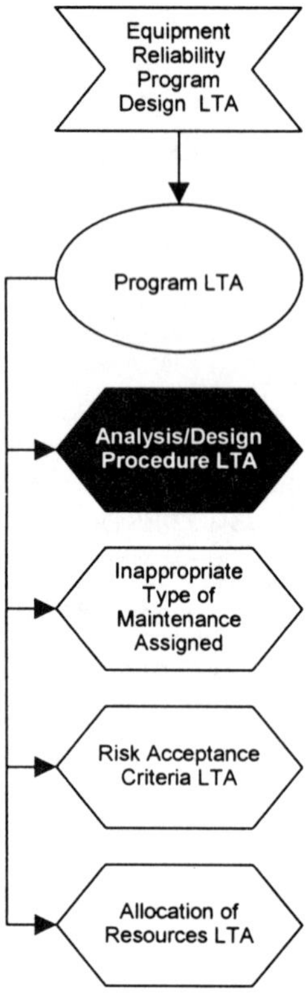

Typical Issues

These include problems related to the design of the maintenance program. Was the process used to determine maintenance tasks completed? Did the process address all equipment important to safety, reliability, and quality? Was the process consistently applied?

Typical Recommendations

- Ensure that the analysis process addresses all aspects of equipment operation important to safety, reliability, and quality
- Ensure that personnel are provided with sufficient guidance to select appropriate maintenance tasks for different types of equipment
- Ensure that personnel who are responsible for developing the equipment reliability program have the proper training

Examples

Maintenance activities had been specified for the running components of a wood chipping machine (i.e., bearings, blades) but no maintenance activities had been specified for the safety interlocks associated with the machine. The analysis procedure did not require safety interlocks to be addressed. As a result, an operator's arm was amputated when it was caught in the chipper and the auto stop feature failed.

Predictive maintenance had been selected for certain conveyors. However, no monitoring program was developed. The equipment reliability analysis program did not require that monitoring programs be developed when predictive maintenance was assigned.

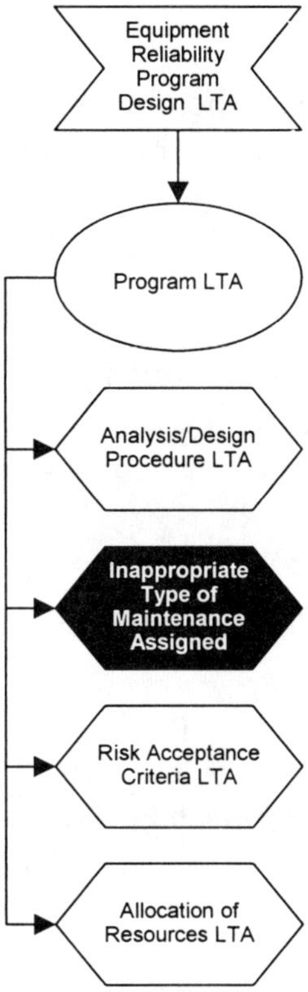

Typical Issues

Was the wrong type of maintenance specified for the equipment? Should corrective maintenance be used instead of proactive maintenance? Should predictive maintenance be assigned instead of proactive maintenance?

Typical Recommendations

- Review equipment failure records to determine if the failures occur at specific intervals of operation or calendar time. Assign preventive maintenance tasks if the risk associated with equipment failure is high enough
- Determine if the failures can be predicted by monitoring a parameter (i.e., pump vibration, temperature, flow). Assign predictive maintenance tasks if the risk associated with equipment failure is high enough
- Determine if failures occur shortly after certain events (i.e., startup, shutdown). Assign proactive maintenance tasks if the risk associated with equipment failure is high enough
- If other types of maintenance are not appropriate, or if the risk associated with the failure is low enough, assign corrective maintenance

Examples

Corrective maintenance was assigned to an auger that provided raw materials to a food process. This selection was based on a very low expected failure rate and a quick repair time. Actual experience indicates the failures took much longer to repair than the analysis team estimated. As a result, the risk associated with the failures was much higher than the team thought.

Records indicated that tube failures were occurring in heat exchangers shortly after plant startup. The failures were determined to be caused by hot spots that developed when contaminants collected in portions of the heat exchanger. Proactive maintenance activities were implemented to clean out the system prior to startup. This removed the contaminants and prevented the heat exchanger failures.

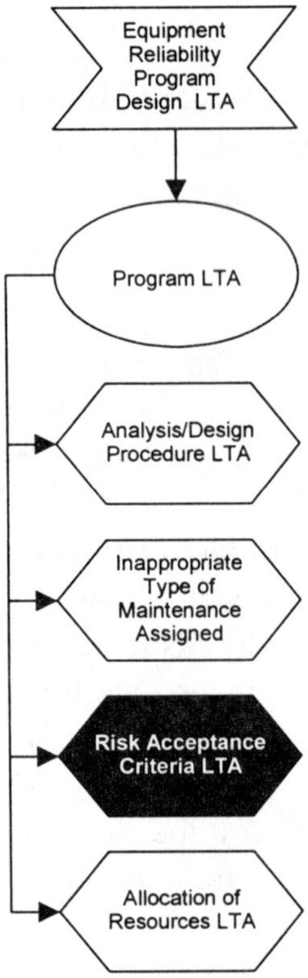

Typical Issues

Were the wrong risk-acceptance criteria used for analyzing the maintenance needs? Was corrective maintenance assigned even though the consequences of failure are very high?

Typical Recommendations

- Ensure that the proper level of risk acceptance is used in determining the level and type of maintenance to perform on equipment
- Provide guidance in the analysis procedure to allow consistent assessment of risk
- Provide guidance in the analysis procedure to allow for consistent application of the risk acceptance criteria. Use specific examples

Examples

Corrective maintenance was assigned to a conveyor that provided raw materials to a food process. Experience indicated that failures took about 16 hours to repair. The analysis procedure did not consider repair times in the overall risk associated with a failure.

The analysis team assigned predictive, proactive, and preventive maintenance activities to equipment with failures that resulted in large consequences. They assigned corrective maintenance to equipment with failures that had only low consequences. However, the risk associated with the low consequence, high frequency events was larger than that associated with some of the high consequence, infrequent events. The risk acceptance criteria outlined in the analysis procedure led them to believe that they were assigning the correct type of maintenance to these different types of risks.

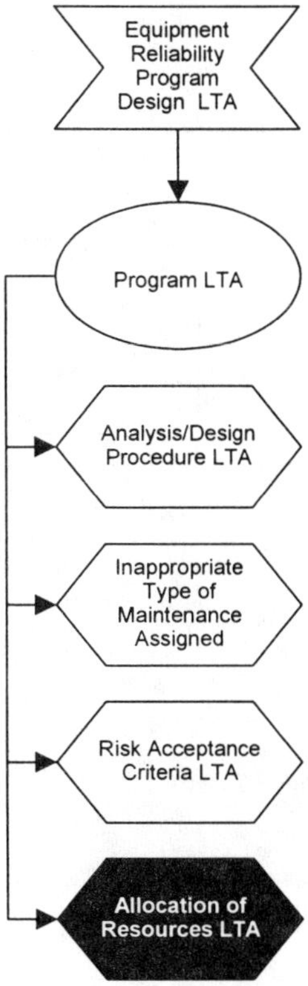

Typical Issues

Are resources assigned based on the risk analysis? Are some high priority tasks not being accomplished because other low priority tasks are being implemented instead?

Typical Recommendations

- Ensure that resources are assigned in accordance with the priorities determined in the equipment reliability program analysis
- Assign additional resources to equipment with a demonstrated history of problems
- Reduce maintenance on equipment that has no significant impact on safety or production and that can be easily repaired or replaced

Examples

Maintenance activities were being conducted for the running components of the wood chipping machine (i.e., bearings, blades) but no maintenance activities were being implemented for the safety interlocks associated with the machine. The equipment reliability program required weekly checks of the interlocks.

Mechanics were always being pulled from scheduled work to work on "emergencies." The percentage of corrective maintenance was 80%. This had not changed since the development of additional preventive, predictive, and proactive maintenance activities.

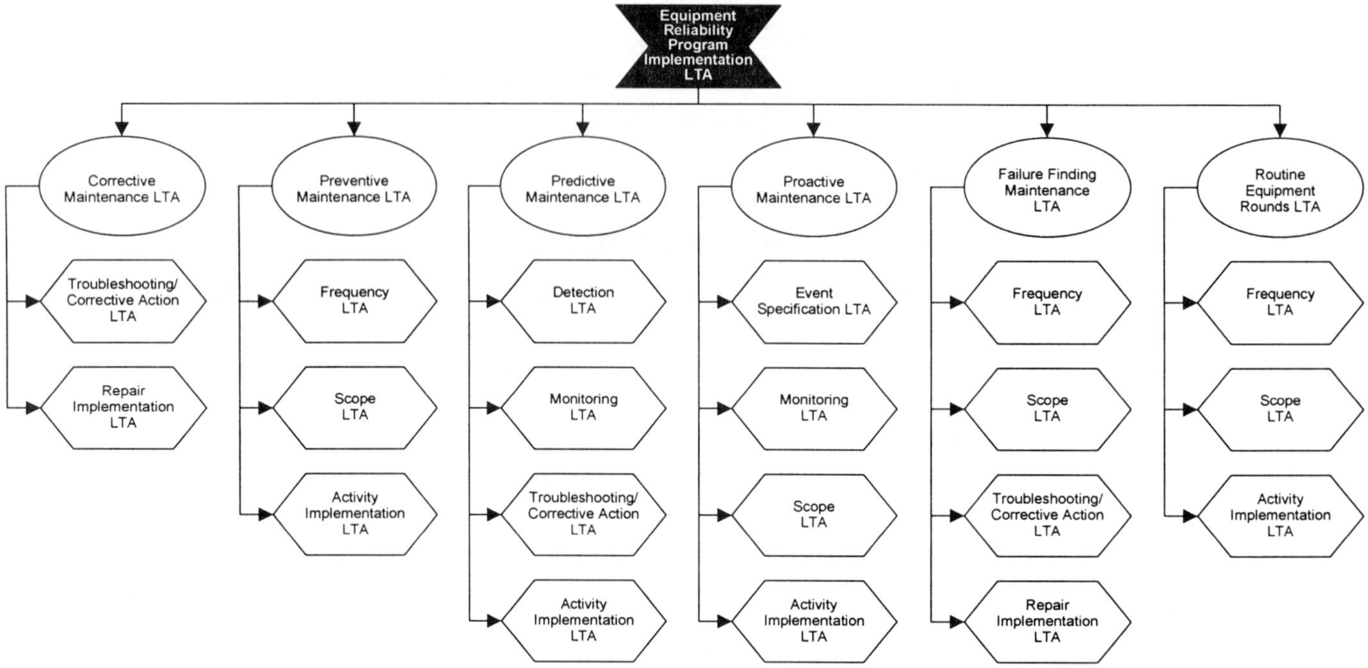

Typical Issues

These include problems related to the implementation of maintenance activities. Was the repair incorrectly performed? Was the troubleshooting less than adequate? Did the monitoring activity fail to detect a failing component? Was maintenance performed when it should have been (i.e., following a shutdown, before a startup, when vibration readings reached a trigger point)?

Typical Recommendations

- Provide troubleshooting guides based on equipment failure analyses for diagnosis of failed components
- Review the frequency of preventive maintenance. If the same activity routinely needs to be performed between scheduled intervals, shorten the preventive maintenance interval
- Ensure that equipment monitoring for predictive maintenance is appropriate for the component

Example

A number of pump bearings have failed recently. Predictive maintenance was selected as the appropriate type of maintenance for the pump bearings. However, monitoring of the pump bearings was never performed even though it was identified as a requirement in the equipment reliability program. As a result, the pump failed before the predictive maintenance activity was implemented.

Preventive maintenance (a calibration) was being performed on a product scale every 3 months. However, operators requested additional calibrations about once per month as they noticed the scale drifting. The frequency of the calibration was changed to once per month after the company was fined for shipping overloaded trucks.

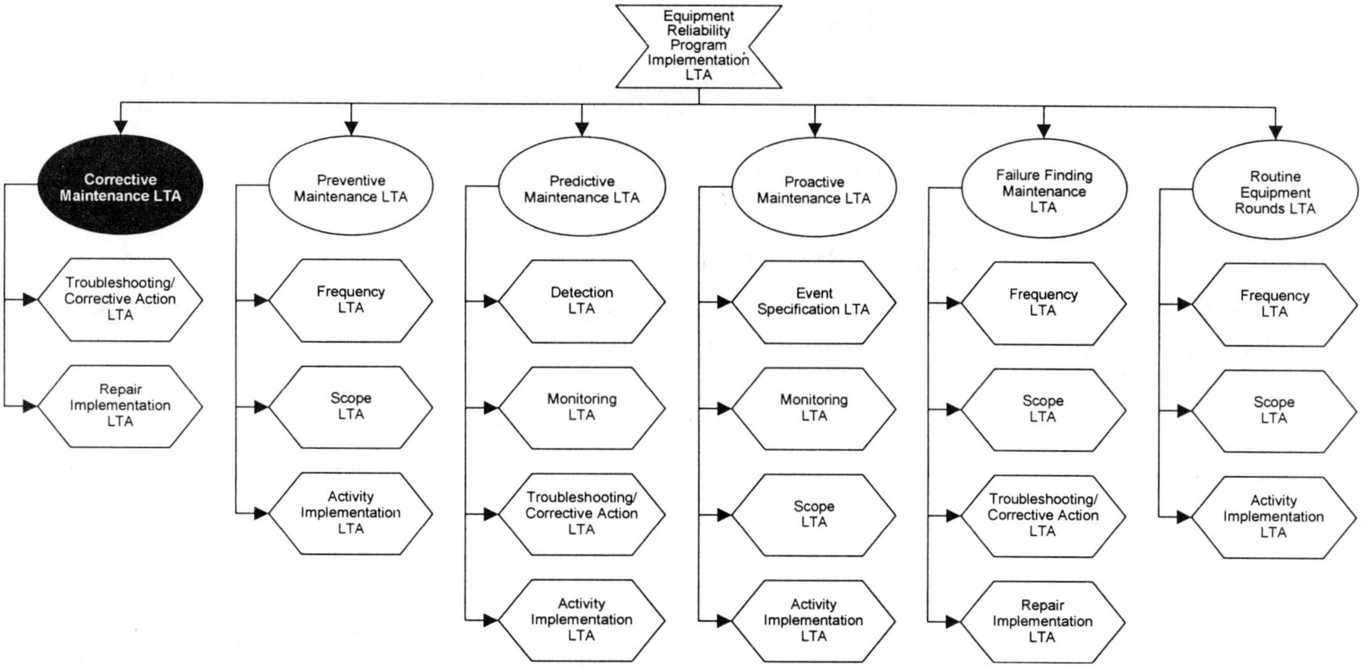

Typical Issues

Was the problem misdiagnosed? Was the corrective maintenance repair performed correctly?

Note: *Dual coding under* Training *or* Procedures *may also be appropriate.*

Typical Recommendations

- Provide troubleshooting guides based on equipment failure analyses for diagnosis of failed components
- Review maintenance procedures to ensure that they provide adequate guidance based on the experience level of personnel
- Provide training for personnel on troubleshooting processes
- Provide training for personnel on repair techniques
- Perform post-maintenance testing to ensure that the maintenance is properly performed and corrects the problem

Examples

Mechanics' job performance was judged by how may work requests they completed. As a result, they tried to diagnose the problem as quickly as possible. This led to rework when the original repairs failed to correct the problem.

An inexperienced mechanic incorrectly repaired a pump seal, which subsequently leaked. He inserted one of the rubber seals backwards. The procedure provided no guidance other than to "install the rubber seals."

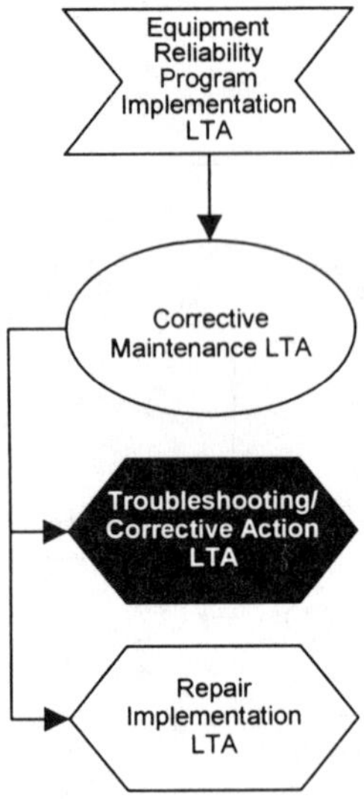

Typical Issues

Was the problem misdiagnosed? Was the wrong problem corrected because the troubleshooting was less than adequate?

Note: *Dual coding under* Training *or* Procedures *may also be appropriate.*

Typical Recommendations

- Provide troubleshooting guides based on equipment failure analyses for diagnosis of failed components
- Provide training for personnel on troubleshooting processes
- Perform post-maintenance testing to ensure the maintenance is properly performed and corrects the problem.

Examples

Mechanics' job performance was judged by how may work requests they completed. As a result, they tried to diagnose the problem as quickly as possible. This led to rework when the original repairs failed to correct the problem.

The electricians were attempting to isolate a ground in a feeder circuit. They thought they had isolated the problem to a portion of the circuit, but they were mistaken. They had misread the electrical diagrams and misinterpreted their instrument readings.

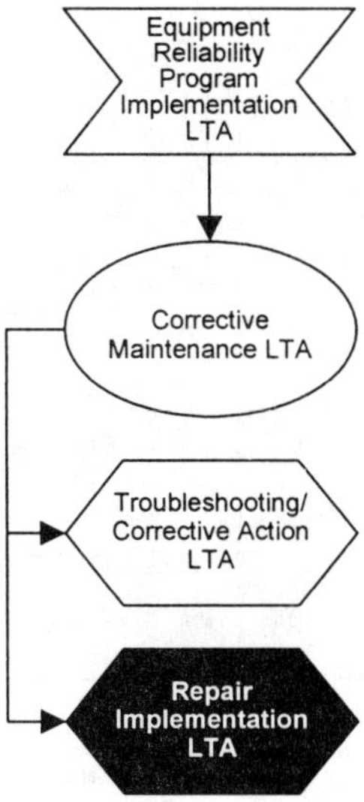

Typical Issue

Was the corrective maintenance repair performed correctly?

Note: *Dual coding under* Training *or* Procedures *may also be appropriate.*

Typical Recommendations

- Review maintenance procedures to ensure that they provide adequate guidance based on the experience level of personnel
- Provide training for personnel on repair techniques

Examples

An inexperienced mechanic incorrectly repaired a pump seal, which subsequently leaked. He inserted one of the rubber seals backwards. The procedure provided no guidance other than to "install the rubber seals."

During corrective maintenance, mechanics identified a problem with a seal on a pressure transmitter. To correct the problem, a new rubber gasket should have been installed. However, the mechanic would have had to go to the warehouse to get a new gasket and it was close to quitting time. Instead, the operator applied a sealant to the gasket. This caused problems during subsequent repairs when the old seal could not be removed.

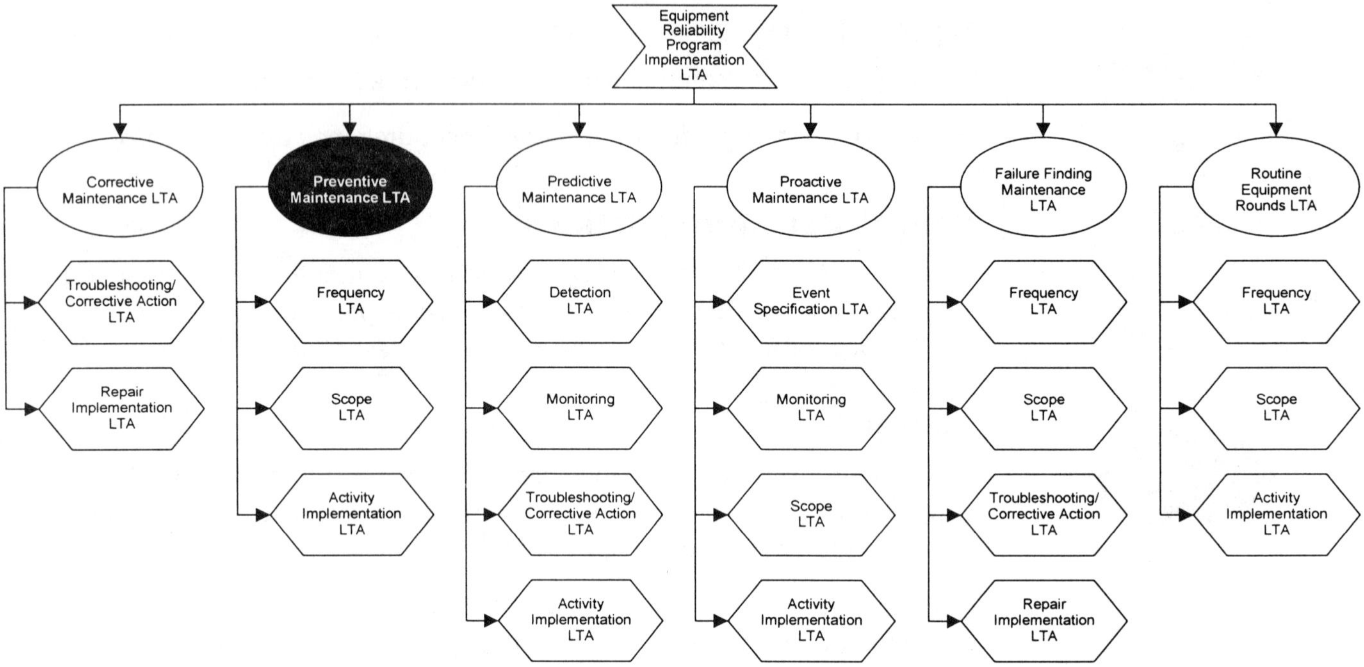

Typical Issues

Was the frequency of the preventive maintenance correct (i.e., too long or too short)? Was the scope of the preventive maintenance activity appropriate (i.e., too broad or too narrow)? Was the activity incorrectly performed?

Typical Recommendations

- Review the frequency of preventive maintenance. If the same activity routinely needs to be performed between scheduled intervals, shorten the preventive maintenance interval
- Review maintenance procedures to ensure that they provide adequate guidance based on the experience level of personnel
- Provide training for personnel on preventive maintenance techniques

Examples

Preventive maintenance (a calibration) was being performed on a product scale every 3 months. However, operators requested additional calibrations about once per month as they noticed the scale drifting. The frequency of the calibration was changed to once per month after the company was fined for shipping overloaded trucks.

Preventive maintenance was being performed on a furnace every week to prevent a buildup of powdered material. However, only the main chamber was being cleaned. Portions of the furnace were not being cleaned, and, as a result, the performance of the furnace degraded over time.

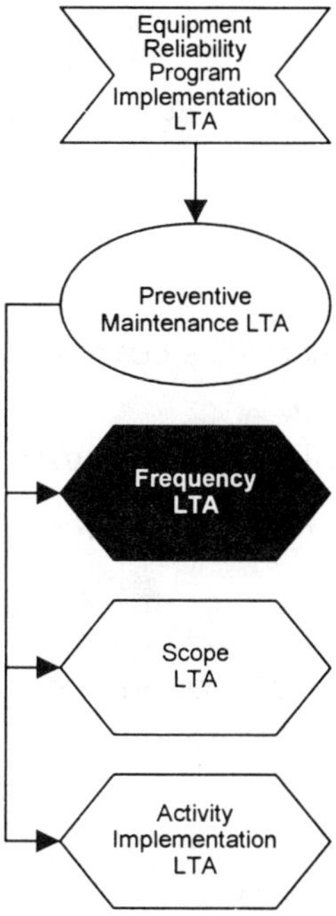

Typical Issue

Was the frequency of the preventive maintenance correct (i.e., too often or not often enough)?

Typical Recommendations

- Review the frequency of preventive maintenance. If the same activity routinely needs to be performed between scheduled intervals, shorten the preventive maintenance interval
- Review the frequency of preventive maintenance. Consider reducing the frequency of preventive maintenance on components. Monitor equipment performance to determine the effects of a reduced frequency

Examples

Preventive maintenance (a calibration) was being performed on a product scale every 3 months. However, operators requested additional calibrations about once per month as they noticed the scale drifting. The frequency of the calibration was changed to once per month after the company was fined for shipping overloaded trucks.

Preventive maintenance was being performed every month on conveyor #1. The maintenance took 6 hours and accounted for 50% of the conveyor's unavailability. Conveyor #2 in a similar service in another part of the plant had the same preventive maintenance performed every 6 months. No failures had occurred on either conveyor in the past 3 years. The preventive maintenance interval for conveyor #1 was changed to once every 6 months.

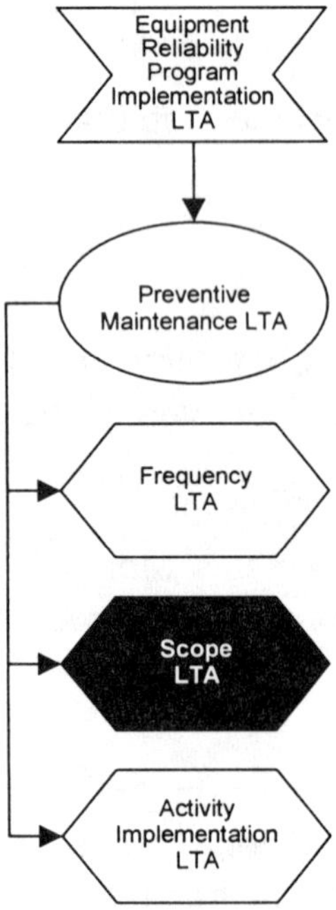

Typical Issue

Was the scope of the preventive maintenance activity appropriate (i.e., too broad or too narrow)?

Typical Recommendations

- Ensure that the scope of preventive maintenance activities covers all portions of the equipment that need repair or service
- Ensure that all of the components requiring preventive maintenance are covered by the procedures

Examples

Preventive maintenance procedures require heavy pieces of inoperative rotating equipment that are not in operation to be rotated to prevent the shafts from warping. Equipment that is shut down is scheduled to be rotated once per week. However, equipment in the warehouse is not covered by the procedure. As a result, some heavy rotors fail after installation.

Preventive maintenance was being performed on a furnace every week to prevent a buildup of powdered material. However, only the main chamber was being cleaned. Portions of the furnace were not being cleaned, and, as a result, the performance of the furnace degraded over time.

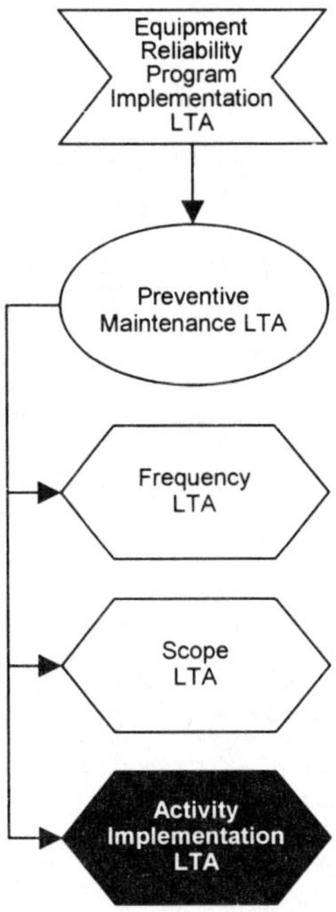

Typical Issues

Was the preventive maintenance activity incorrectly performed? Were all required components serviced? Were some items included on the schedule that were never performed?

Note: *Dual coding under* Training, Procedures, *or* Planning, Scheduling, or Tracking of Work Activities LTA (Administrative/Management Systems, SPACs LTA) *may also be appropriate.*

Typical Recommendations

- Review maintenance procedures to ensure that they provide adequate guidance based on the experience level of personnel
- Provide training for personnel on preventive maintenance techniques
- Review the preventive maintenance schedule and completed work orders to ensure that all required activities are being performed
- Perform post-maintenance testing to ensure that the maintenance is properly performed

Examples

An inexperienced mechanic incorrectly installed a pump seal, which subsequently leaked. He inserted one of the rubber seals backwards. The procedure provided no guidance other than to "install the rubber seals."

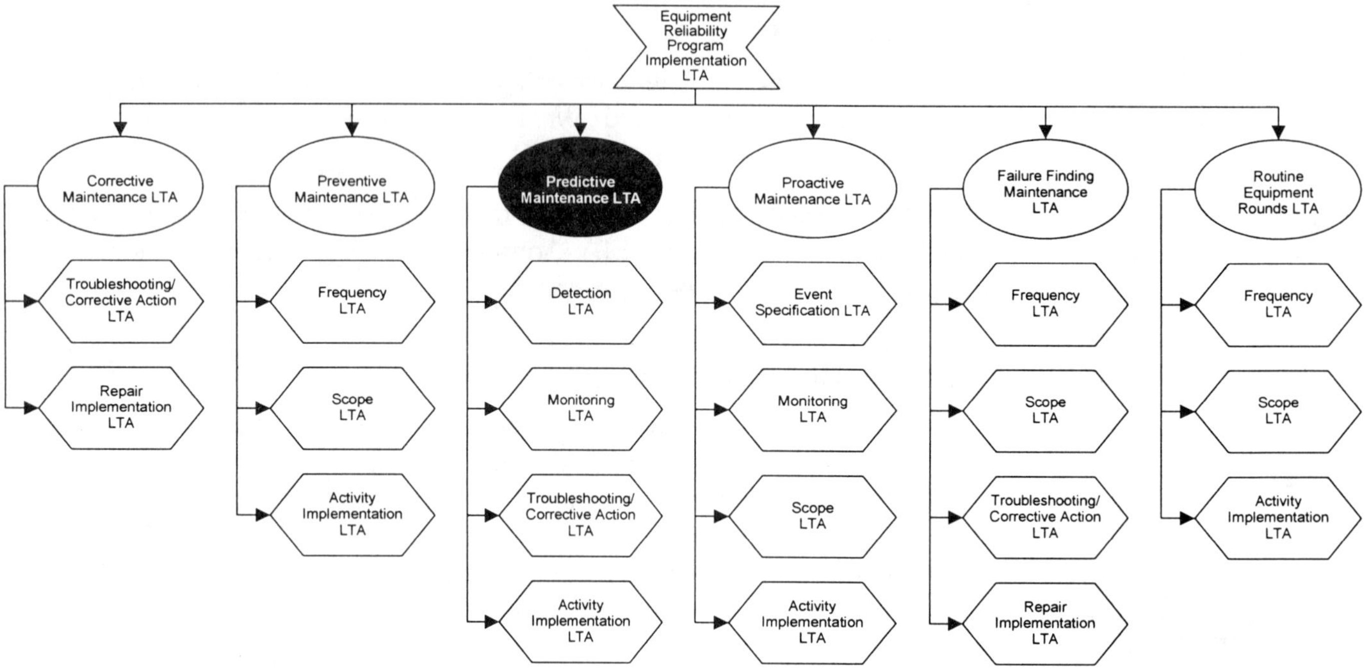

Typical Issues

Did the monitoring activity fail to detect a failing component? Was the monitoring activity performed? Was the correct parameter being monitored to detect failure? Was the predictive maintenance incorrectly performed?

Typical Recommendations

- Provide guidance on the typical parameters that can be monitored to predict failures for different types of components
- Ensure that equipment monitoring for predictive maintenance is appropriate for the component
- Ensure that equipment monitoring is being performed
- Ensure that the scope of equipment monitoring is adequate

Examples

A number of pump bearings have failed recently. Predictive maintenance was selected as the appropriate type of maintenance for the pump bearings. However, monitoring of the pump bearings was never performed even though it was identified as a requirement in the equipment reliability program. As a result, the pump failed before the predictive maintenance activity was implemented.

Monitoring of a pump indicated an upcoming failure (e.g., from predictive maintenance monitoring). The pump was repaired incorrectly.

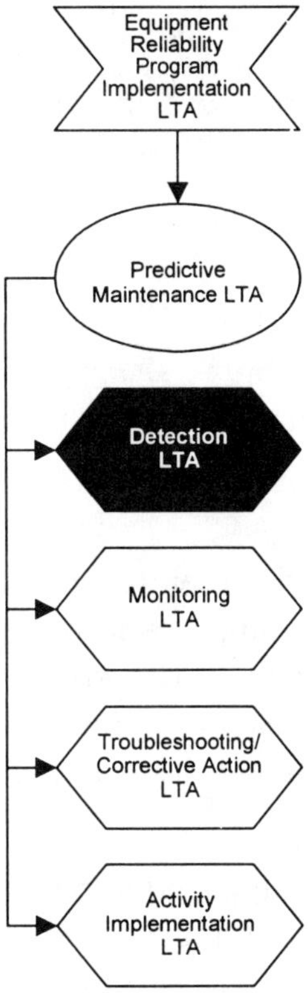

Typical Issues

Did the monitoring activity fail to detect a failing component? Was the correct parameter being monitored to detect failure? Is there sufficient time to detect an impending failure before the failure actually occurs?

Typical Recommendations

- Provide guidance on the typical parameters that can be monitored to predict failures for different types of components
- Ensure that equipment monitoring for predictive maintenance is appropriate for the component

Examples

Pump bearings were being monitored for failure. However, by the time the impending failure could be detected, there was insufficient time to perform the maintenance.

Turbine bearing temperatures were being monitored to predict impending failures. However, failures occurred even though there was no prediction of failure based on temperature levels. Vibration should have been monitored instead, because it was a better predictor of impending failures.

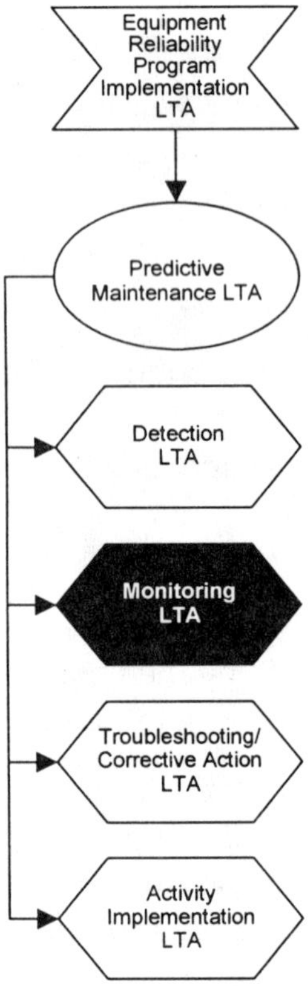

Typical Issues

Were monitoring activities being performed? Were all equipment and all components monitored?

Typical Recommendations

- Ensure that equipment monitoring is being performed
- Ensure that all pieces of equipment are being monitored
- Ensure that all components (points) are being monitored

Examples

A number of pump bearings have failed recently. Predictive maintenance was selected as the appropriate type of maintenance for the pump bearings. However, monitoring of the pump bearings was never performed even though it was identified as a requirement in the equipment reliability program. As a result, the pump failed before the predictive maintenance activity was implemented.

The three supply fans for the assembly building were all supposed to be monitored for vibration as part of predictive maintenance. Only two of the three fans were being monitored. The third fan was difficult to access.

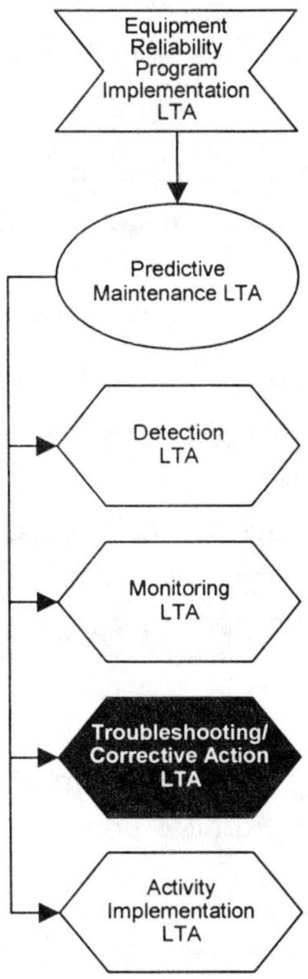

Typical Issues

Was the scope of the work appropriate? Did the maintenance address the problem? Was the scope broad enough to correct the problem?

Note: *Dual coding under* Training *or* Procedures *may be appropriate.*

Typical Recommendations

- Provide guidance on the typical failures that occur in various components
- Provide troubleshooting guides based on equipment failure analyses for diagnosis of failed components
- Provide training for personnel on troubleshooting processes
- Perform post-maintenance testing to ensure that the maintenance is properly performed and that it corrects the problem

Examples

Mechanics' job performance was judged by how many work requests they completed. As a result, they tried to diagnose the problem as quickly as possible. This led to rework when the original repairs failed to correct the problem.

High vibration readings generally indicated a bearing problem in the pump. The mechanics replaced the bearing even though it did not look worn or damaged. When the pump was restarted, the high vibration readings were still present. The pump impeller had been damaged and caused the high vibration. This was not considered as a potential cause of the high vibration.

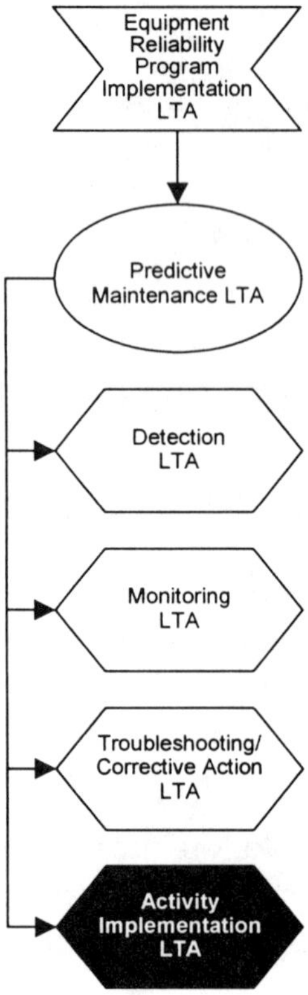

Typical Issue

Was the predictive maintenance incorrectly performed?

Note: *Dual coding under* Training, Procedures, *or* Planning, Scheduling, or Tracking of Work Activities LTA (Administrative/Management Systems, SPACs LTA) *may also be appropriate.*

Typical Recommendations

- Review maintenance procedures to ensure that they provide adequate guidance based on the experience level of personnel
- Provide training for personnel on predictive maintenance techniques
- Review the preventive maintenance schedule and completed work orders to ensure that all required activities are being performed

Example

An inexperienced mechanic incorrectly installed a pump seal. He inserted one of the rubber seals backwards. The procedure provided no guidance other than to "install the rubber seals."

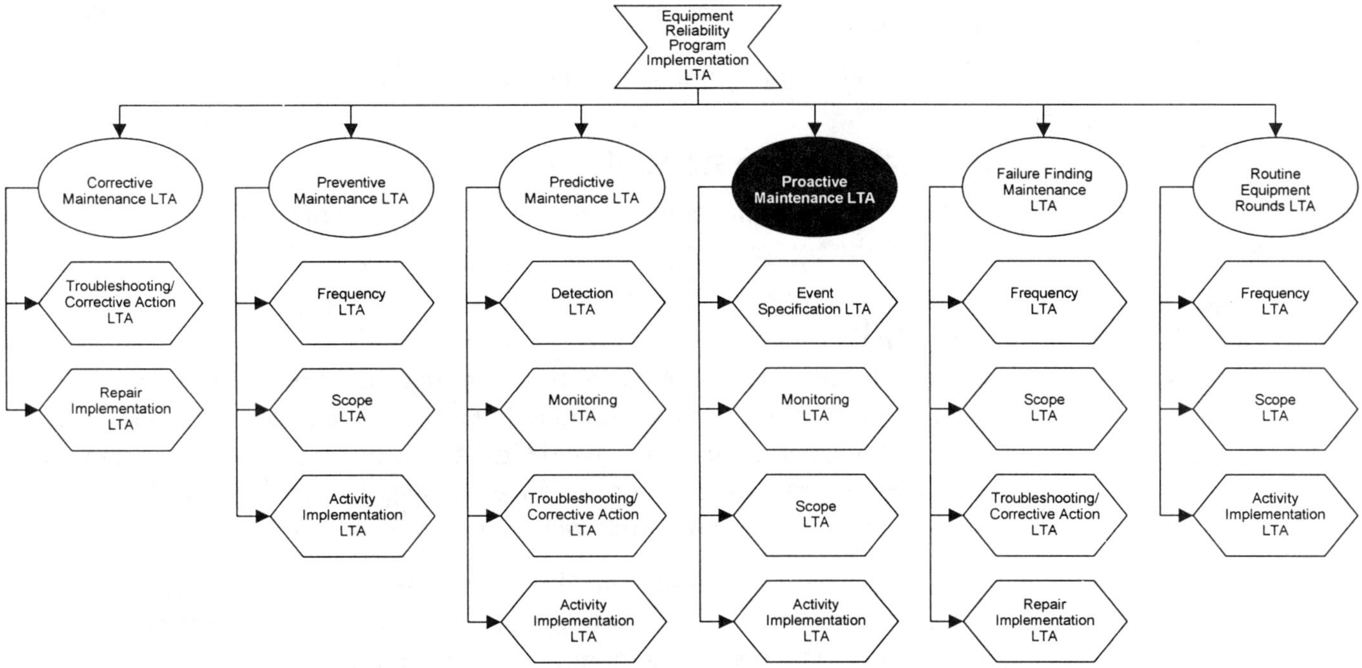

Typical Issues

Was maintenance performed when it should have been (i.e., following a shutdown, before a startup, the beginning of winter)? Was the work incorrectly performed? Was the scope of the activity broad enough?

Typical Recommendations

- Ensure that triggering events for proactive maintenance are appropriate for the component
- Ensure that monitoring is performed to determine when triggering events occur
- Review maintenance procedures to ensure that they provide adequate guidance based on the experience level of personnel
- Provide training for personnel on monitoring and maintenance techniques
- Review the proactive maintenance schedule and completed work orders to ensure that all required activities are being performed

Examples

Product barrels were cleaned as they were returned from customers. However, some product was contaminated by dust that accumulated in those barrels that were not used for an extended period of time. Cleaning was switched to shortly before use instead of when the barrels were returned from customers.

Cranes were supposed to be inspected and lift-tested prior to lifting any item that was greater than 70% of the crane's rated capacity. These inspections and tests were never performed because the crane operators were unaware of this requirement.

Furnace crucibles were to be cleaned whenever the furnace was scheduled to be shut down for more than 8 hours. Operations never told maintenance when the scheduled shutdowns would occur. As a result, the cleaning was not performed as required.

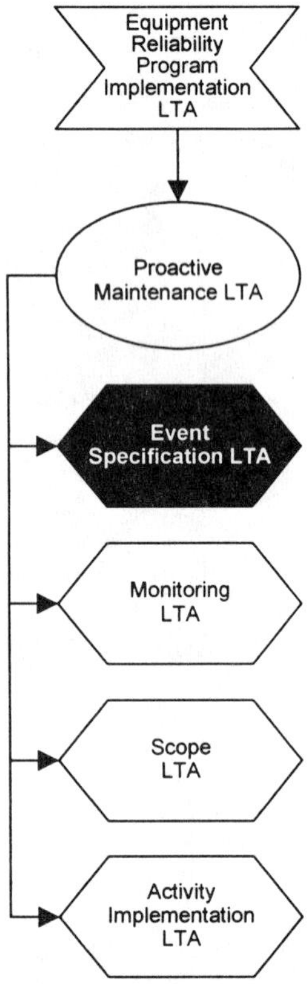

Typical Issue

Is the correct triggering event specified for the proactive maintenance?

Typical Recommendations

- Ensure that triggering events for proactive maintenance are appropriate for the component

Example

Product barrels were cleaned as they were returned from customers. However, some product was contaminated when it was placed in the barrels. In barrels that were not used for extended periods of time, dust would accumulate and contaminate the product. Cleaning was switched to shortly before use instead of when the barrels were returned from customers.

Tubes in a heat exchanger were failing prematurely. The tubes were rinsed prior to starting a new batch, but they were not cleaned at the completion of each batch. As a result, the material remaining in the tubes caused the tubes to corrode.

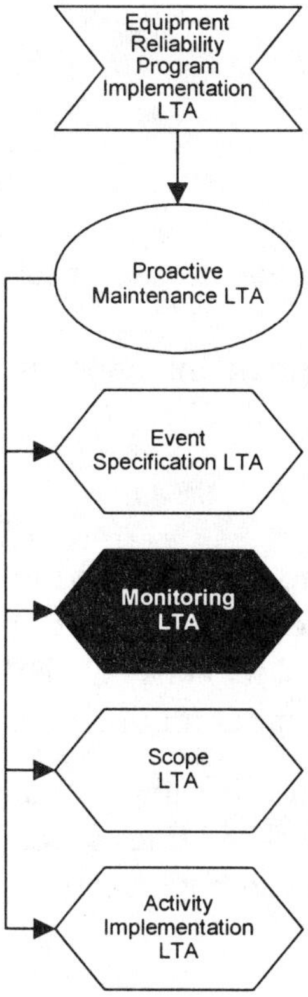

Typical Issues

Was a monitoring program in place to determine when these events occurred? Was maintenance notified when these events occurred?

Typical Recommendations

- Ensure that monitoring is performed to determine when triggering events occur
- Review the proactive maintenance schedule and completed work orders to ensure that all required activities are being performed

Examples

Cranes were supposed to be inspected and lift tested prior to lifting any item that was greater than 70% of the crane's rated capacity. These inspections and tests were never performed because the crane operators were unaware of this requirement.

Furnace crucibles were to be cleaned whenever the furnace was scheduled to be shut down for more than 8 hours. Operations never told maintenance when the scheduled shutdowns would occur. As a result, the cleaning was not performed as required.

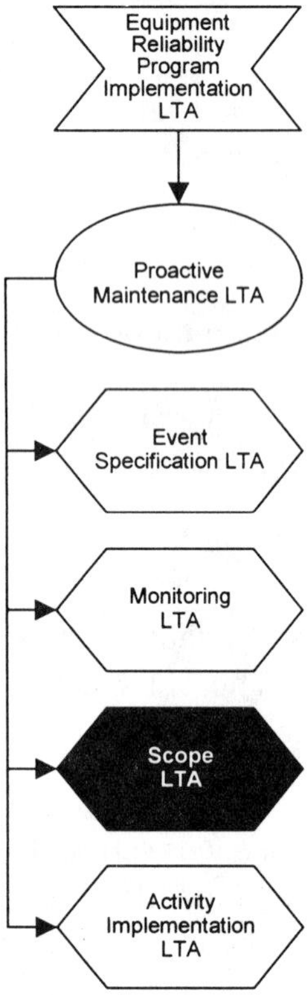

Typical Issues

Was the scope of the activity sufficient to prevent the problem?

Typical Recommendations

- Review the scope of the proactive maintenance procedures to ensure that they are broad enough to address the issue
- Perform post-maintenance testing to ensure that the maintenance is properly performed and corrects the problem

Example

At the end of the season, lawn mowers were supposed to be winterized to prevent damage while sitting idle over the winter. Maintenance changed the oil, but failed to stabilize the gas. As a result, the mowers' fuel lines were gummed up in the spring when the mowers were brought out for use.

Cranes were supposed to be inspected and lift tested prior to lifting any item that was greater than 70% of the crane's rated capacity. The lift tests were only performed at one boom angle even though they should have been performed at a number of different boom angles.

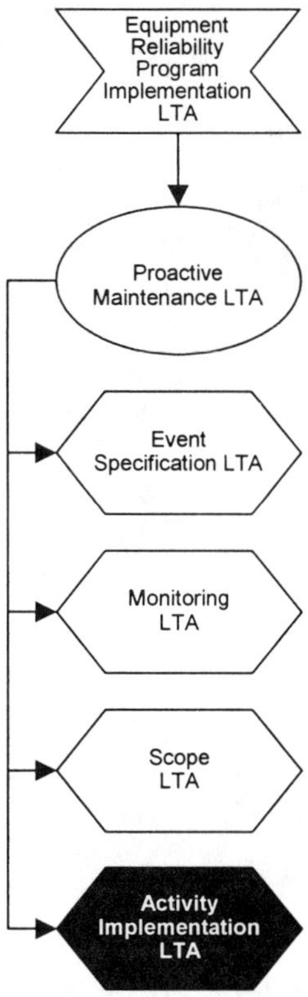

Typical Issue

Was the maintenance incorrectly performed?

Note: *Dual coding under* Training, Procedures, *or* Planning, Scheduling, or Tracking of Work Activities LTA (Administrative/Management Systems, SPACs LTA) *may also be appropriate.*

Typical Recommendations

- Review maintenance procedures to ensure that they provide adequate guidance based on the experience level of personnel
- Provide training for personnel on repair techniques
- Review the proactive maintenance schedule and completed work orders to ensure that all required activities are being performed

Examples

An inexperienced mechanic incorrectly installed a pump seal. He inserted one of the rubber seals backwards. The procedure provided no guidance other than to "install the rubber seals."

An electrician was performing a maintenance check on a pressure instrument. During performance of the check, a high pressure signal was simulated in the instrument loop. Because the loop was not properly isolated, it resulted in a pressure relief valve lifting and a release to the environment.

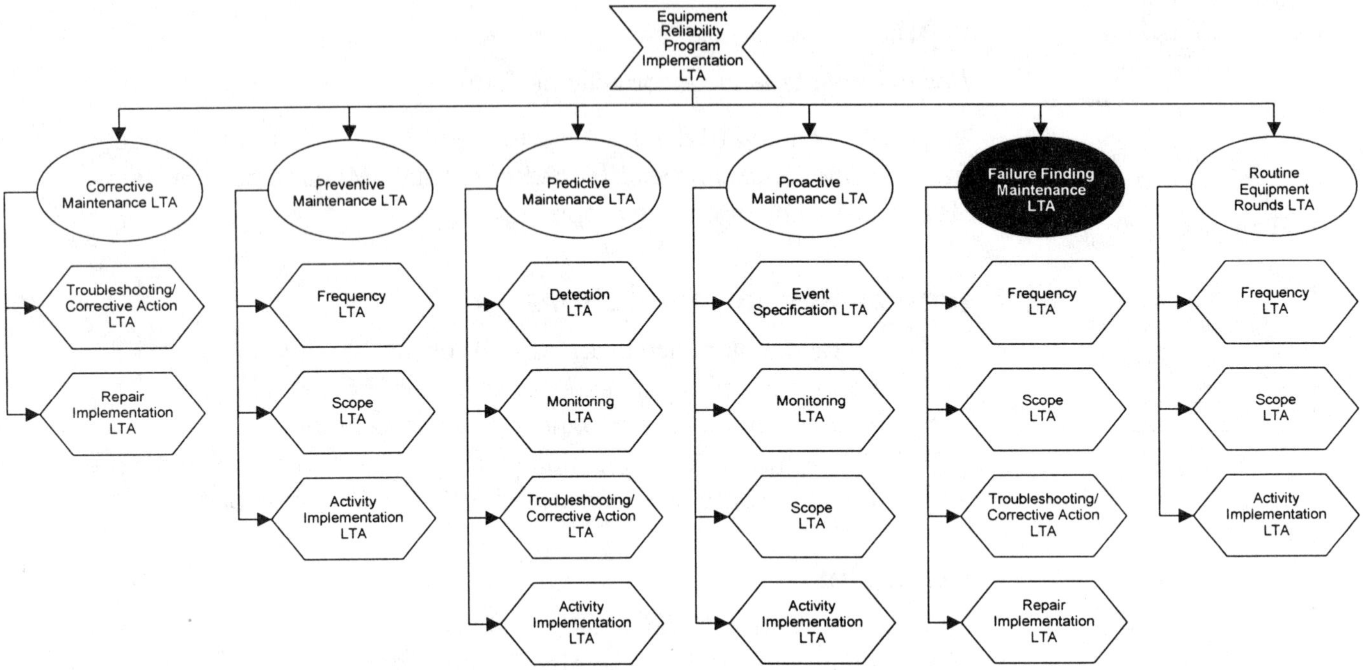

Typical Issues

Did hidden failures contribute to the loss event? Could these hidden failures have been detected by testing the equipment?

Note: *This type of maintenance is usually applicable to standby systems or the detection of hidden failures in systems.*

Typical Recommendations

- Ensure that standby systems are periodically tested to determine their operability
- Verify that installed spares are periodically used to ensure that they are ready to operate when the primary components or trains fail
- Check fault-finding testing procedures to ensure that they test the entire system and not just a portion of it
- Ensure that the frequency of testing is correct (not too often, but often enough)

Examples

A standby diesel generator was installed to provide power to vital components during a loss of power. No testing had been performed on the diesel generator since it was installed. As a result, when there was a loss of power, the diesel generator did not work.

A second cooling pump is installed as a spare. It is designed to start when the primary pump fails. The standby pump is smaller than the primary and so it is seldom used. The pump is tested when it is periodically placed in service (although this is not done on any schedule). However, the autostart system is never tested. As a result, the standby pump failed to start following an emergency shutdown of the primary pump.

Routine testing of a computer backup power supply (an uninterruptible power supply with batteries) was performed once a year. However, the batteries had an expected lifetime of 18 months. As a result, many of the battery failures were not detected for months after they occurred.

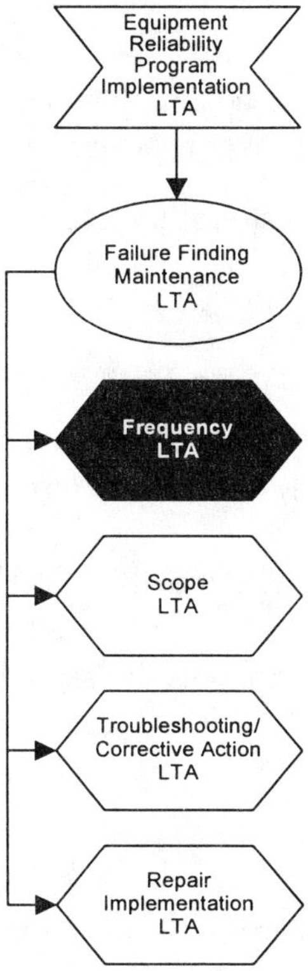

Typical Issues

Was the frequency of fault-finding maintenance correct? Was the maintenance performed too frequently? Was it not performed often enough?

Note: *This type of maintenance is usually applicable to standby systems or the detection of hidden failures in systems.*

Typical Recommendations

- Ensure that standby systems are periodically tested to determine their operability
- Ensure that the frequency of testing is correct (not too often, but often enough)
- Assess the impact of fault-finding maintenance on the system. What impact does the maintenance have on the equipment? Adjust the frequency accordingly

Examples

A standby diesel generator was installed to provide power to vital components during a loss of power. No testing had been performed on the diesel generator since it was installed. As a result, when there was a loss of power, the diesel generator did not work.

Routine testing of a computer backup power supply (an uninterruptible power supply with batteries) was performed once a year. However, the batteries had an expected lifetime of 18 months. As a result, many of the battery failures were not detected for months after they occurred.

An important control system in a nuclear power plant was tested daily to detect hidden failures. The test took about an hour to perform. As a result, the system was inoperable about 5% of the time for scheduled maintenance. In addition, the maintenance often introduced problems into the system that rendered it inoperable. The test frequency was modified so that it was performed once per week.

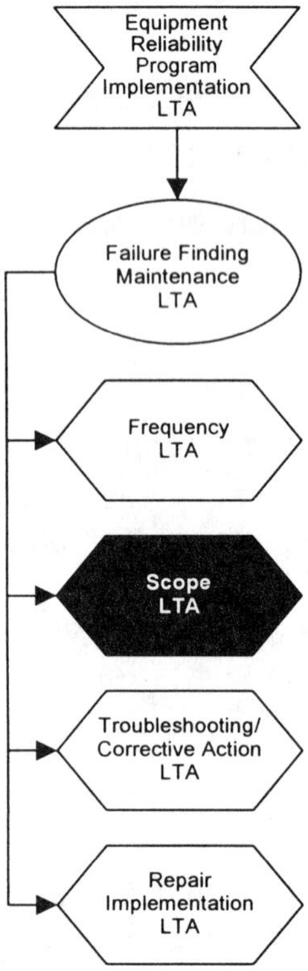

Typical Issues

Did the testing include all applicable portions of the system (i.e., detection system, control systems, actuation systems, and the actual components)?

Note: *This type of maintenance is usually applicable to standby systems or the detection of hidden failures in systems.*

Typical Recommendations

- Check fault-finding testing procedures to ensure that they test the entire system and not just a portion of it. Check to see that the following portions of the system are included:
  - detection systems (i.e., a system that detects low voltage to start an emergency generator)
  - actuation systems (i.e., the part of the system that tells the standby component to start)
  - the component itself (i.e., the diesel generator)

Example

A second cooling pump is installed as a spare. It is designed to start when the primary pump fails. The standby pump is smaller than the primary and so it is seldom used. The pump is tested when it is periodically placed in service (although this is not done on any schedule). However, the autostart system is never tested. As a result, the standby pump failed to start following an emergency shutdown of the primary pump.

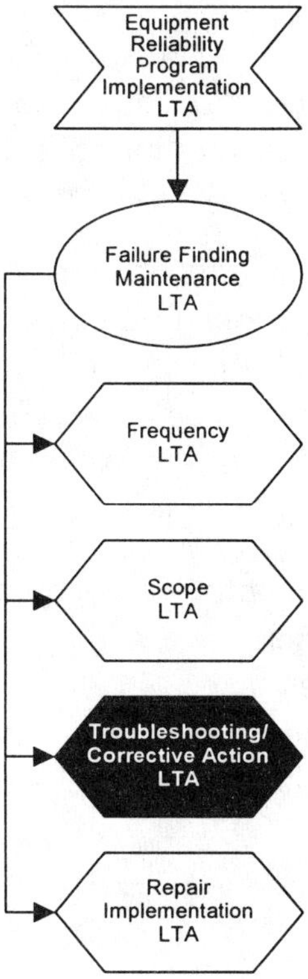

Typical Issues

Was the scope of the repairs appropriate? Did the repair correct the problem? Was the scope of the repair broad enough to correct the problem?

Typical Recommendations

- Provide guidance on the typical failures that occur during testing
- Provide troubleshooting guides based on equipment failure analyses for diagnosis of failed components
- Provide training for personnel on troubleshooting processes
- Perform post-maintenance testing to ensure that maintenance was properly performed and corrects the problem

Example

During testing, a standby generator failed to start. Troubleshooting revealed a failure in the starting circuit. No post-maintenance testing was performed. As a result, a failed fuel line was not discovered.

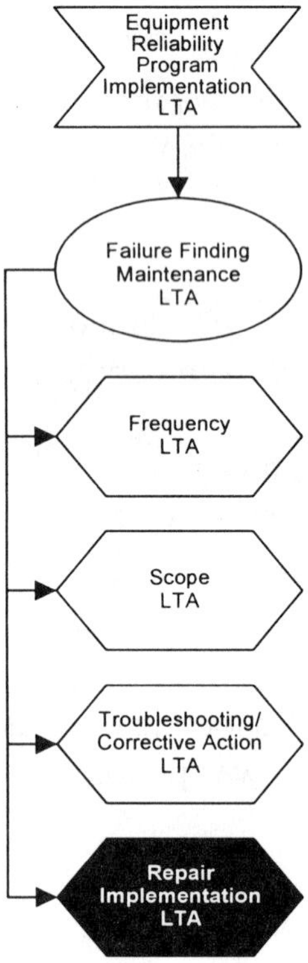

Typical Issues

Was an error made in performing the repair activity? Were problems introduced as a result of performing the repair? Were hidden failures introduced into the system as a result of performing the maintenance?

Note: *This type of maintenance is usually applicable to standby systems or the detection of hidden failures in systems. Dual coding under* Training, Procedures, *or* Planning, Scheduling, or Tracking of Work Activities LTA (Administrative/Management Systems, SPACs LTA) *may also be appropriate.*

Typical Recommendations

- Review maintenance procedures to ensure that they provide adequate guidance based on the experience level of personnel
- Provide training for personnel on repair techniques
- Perform an analysis of procedures to determine the types of errors that could be reasonably made. Ensure that the procedures adequately address each of these

Examples

A standby diesel generator (DG) was installed to provide power to vital components during a loss of power. To perform testing of the DG, the maintenance technician takes the DG off-line. After testing, maintenance failed to return the DG to an on-line condition. As a result, when there was a loss of power, the diesel generator did not work.

A secondary cooling pump is installed as a spare. It is designed to start when the primary pump fails. A failure in the autostarting system was found during a test. However, the pump was not repaired for several weeks because it was not put on the maintenance schedule. When the primary pump tripped, the secondary pump was still inoperable.

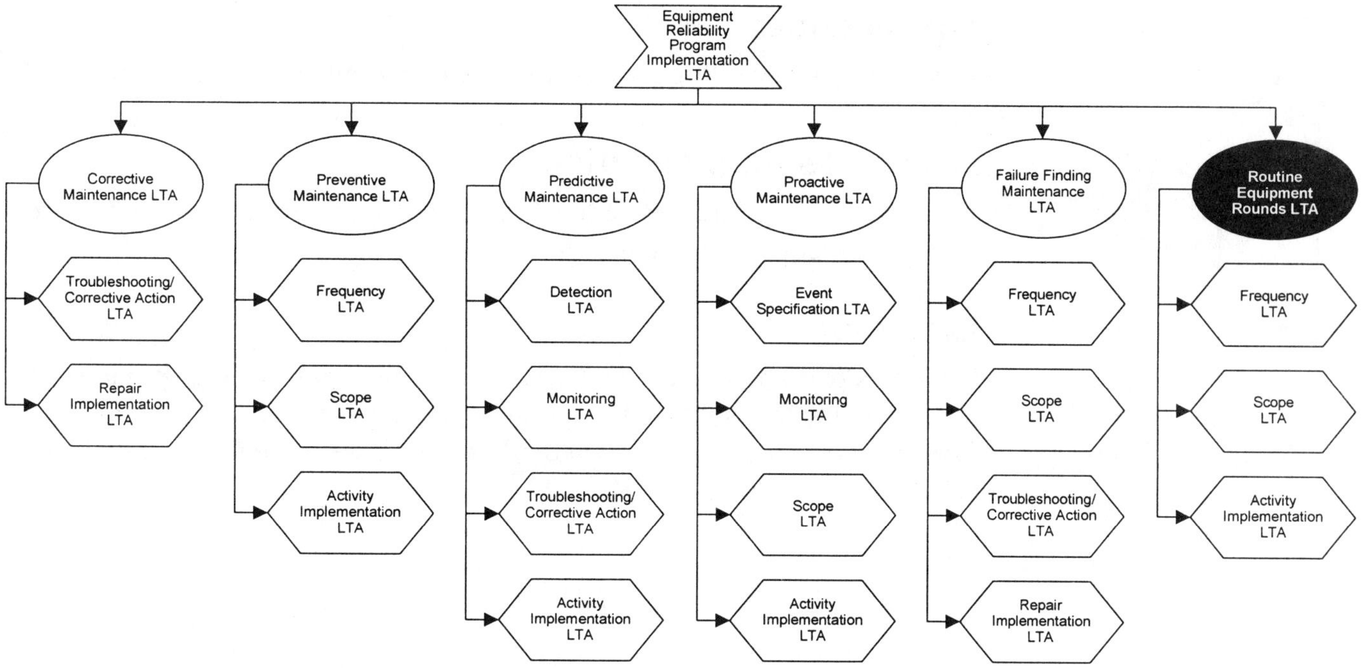

Typical Issues

Are routine inspections of equipment performed? Are personnel aware of the types of problems they should look for? Do they know how to document the problem and feed it into the maintenance system?

Typical Recommendations

- Develop guidance for operator and maintenance rounds
- Ensure that personnel are aware of the process for initiating corrective maintenance
- Make the process of reporting problems as simple as possible to encourage reporting problems

Examples

Operators are supposed to inspect the line for problems at the beginning of each shift. Often the operators skip the rounds because they have too much paperwork to complete.

Grounding straps on a pipeline were to be inspected once per year. Most of the line was inspected yearly, but portions that were difficult to access were frequently skipped.

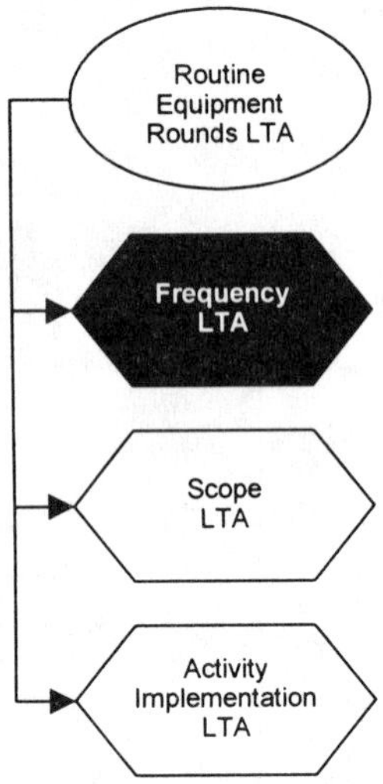

Typical Issues

Was the frequency of the rounds correct (i.e., too often or not often enough)?

Typical Recommendation

- Review the frequency of the rounds to determine if they are performed at the required frequency

Example

Operators performed equipment rounds in some areas of the plant only once a day. Frequently, significant valve packing leaks were found by the operators. More frequent rounds resulted in detections of leaks while they were still very small.

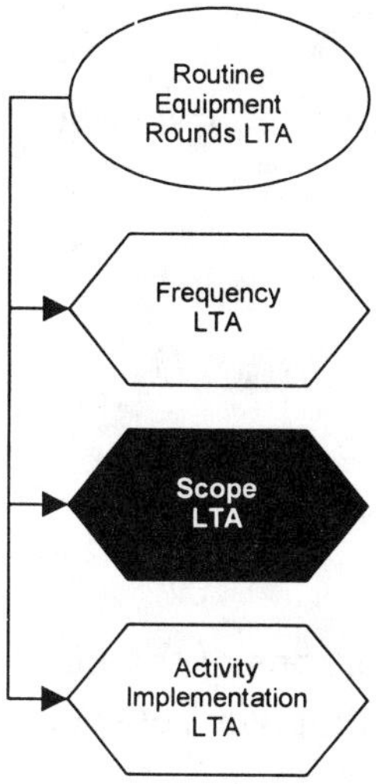

Typical Issues

Was the scope of the rounds appropriate (i.e., too broad or too narrow)? Are all portions of the plant covered by routine rounds?

Typical Recommendations

- Ensure that all areas of the plant are covered by periodic rounds
- Provide guidance on the activities that are to be performed during routine rounds

Examples

Operators were told to perform rounds in the steam plant but were not told what activities they were to perform. As a result, the operators poked their head in the door of the building and glanced around, but did nothing else.

A plant was recently automated. The operators did not need to leave the control room to operate the plant. The operators only toured the area right around the control room. As a result, no one routinely toured the entire plant.

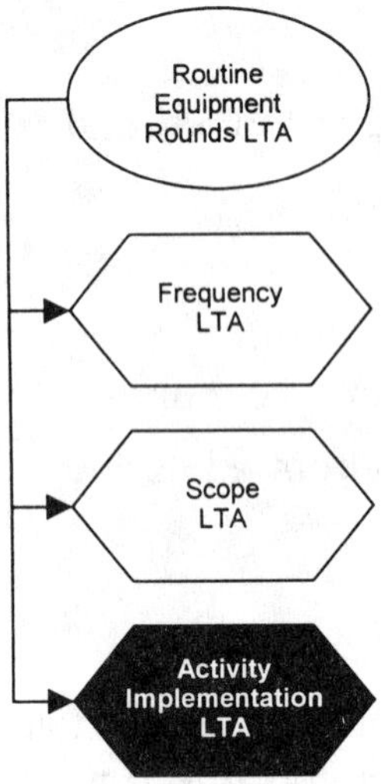

Typical Issues

Are the rounds performed? Are they performed at the specified frequency? Do the rounds cover all areas that are specified?

Typical Recommendations

- Ensure that rounds are performed as required
- Ensure that all equipment is covered on rounds as required

Example

Operators are supposed to check for leaks in various portions of the plant. However, they usually only toured the part of the plant that was between the control room and the lunch room.

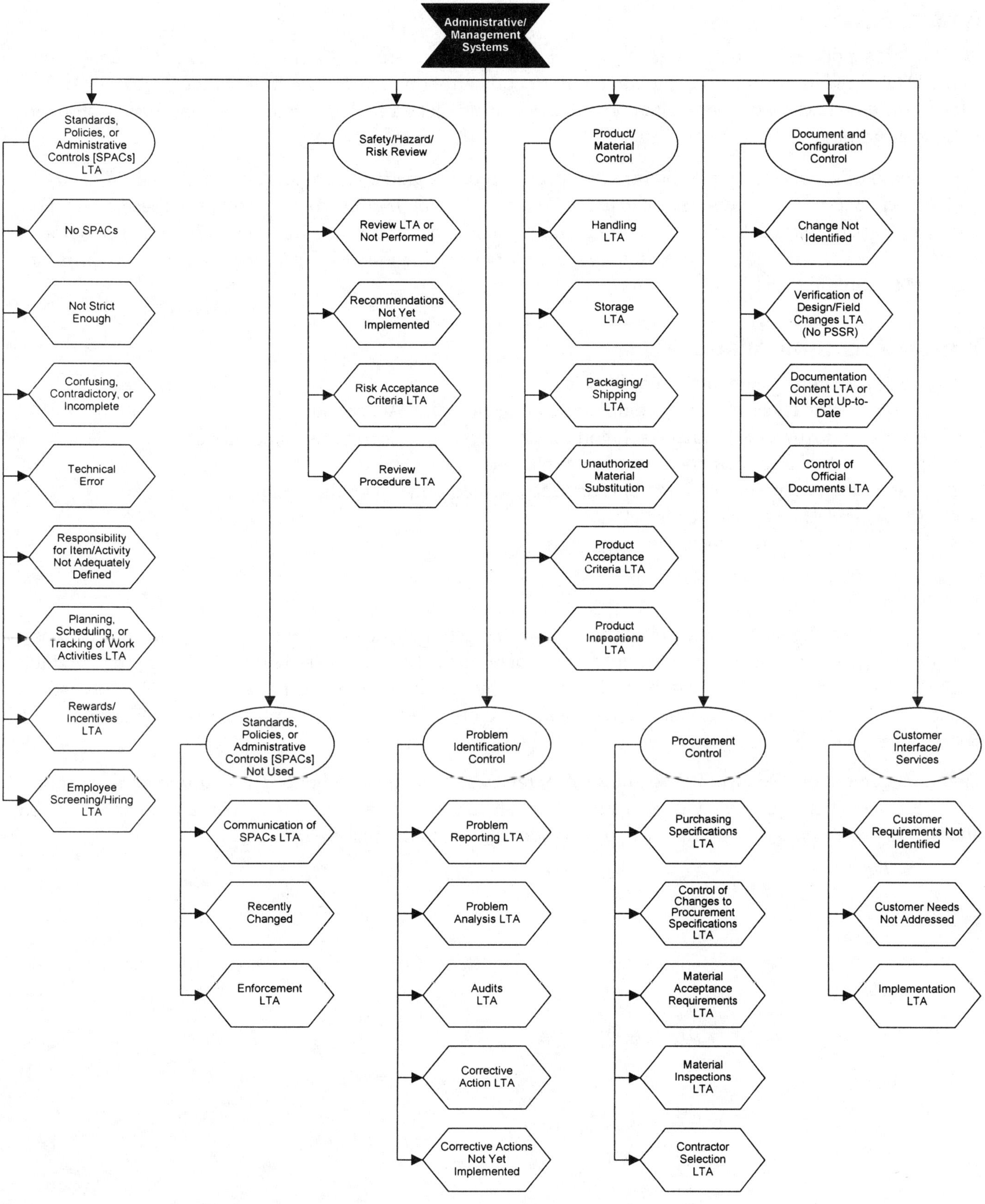
Administrative/ Management Systems
Standards, Policies, or Administrative Controls [SPACs] LTA
No SPACs
Not Strict Enough
Confusing, Contradictory, or Incomplete
Technical Error
Responsibility for Item/Activity Not Adequately Defined
Planning, Scheduling, or Tracking of Work Activities LTA
Rewards/ Incentives LTA
Employee Screening/Hiring LTA
Standards, Policies, or Administrative Controls [SPACs] Not Used
Communication of SPACs LTA
Recently Changed
Enforcement LTA
Safety/Hazard/ Risk Review
Review LTA or Not Performed
Recommendations Not Yet Implemented
Risk Acceptance Criteria LTA
Review Procedure LTA
Problem Identification/ Control
Problem Reporting LTA
Problem Analysis LTA
Audits LTA
Corrective Action LTA
Corrective Actions Not Yet Implemented
Product/ Material Control
Handling LTA
Storage LTA
Packaging/ Shipping LTA
Unauthorized Material Substitution
Product Acceptance Criteria LTA
Product Inspections LTA
Procurement Control
Purchasing Specifications LTA
Control of Changes to Procurement Specifications LTA
Material Acceptance Requirements LTA
Material Inspections LTA
Contractor Selection LTA
Document and Configuration Control
Change Not Identified
Verification of Design/Field Changes LTA (No PSSR)
Documentation Content LTA or Not Kept Up-to-Date
Control of Official Documents LTA
Customer Interface/ Services
Customer Requirements Not Identified
Customer Needs Not Addressed
Implementation LTA

Typical Issues

Do standards, policies, or administrative controls (SPACs) exist? Are they inadequate or inadequately implemented? Did inadequate material, procurement, or configuration control contribute to the problem? Are safety/hazard reviews inadequate? Are corrective actions identified and implemented? Was there a problem with the customer interface or customer service?

Note: *Standards, policies, and administrative controls provide guidance on how an activity should be accomplished, whereas procedures provide a detailed, step-by-step method for performing a specific task. For example, there are SPACs that describe the policies governing scheduling of workers. There is also a procedure that provides a detailed, step-by-step process for performing the task, including the forms to complete and data to enter in the computer system.*

Typical Recommendations

- Provide written documentation of SPACs
- Ensure that all levels of affected employees are aware of SPAC changes
- Track and document the final resolution for all corrective action recommendations
- Inspect materials for damage upon arrival at the facility
- Ensure that acceptance requirements are documented and match the design requirements
- Review and approve field changes
- Periodically solicit feedback from customers

Examples

A mechanic installing a cable tray drilled into a live wire within a wall because the facility drawings he was using were not up-to-date. A management system for control of electrical drawings may have prevented this occurrence by ensuring that the mechanic had up-to-date documentation. A management policy/procedure would also be required to ensure that such drawings are obtained/reviewed as part of the work permit system for penetrations of any wall.

The management of change policy requires safety reviews of all process changes, but during an overnight emergency, a failed gate valve was replaced with a ball valve. The hazards of the change were not reviewed, and the valve subsequently ruptured when peroxide trapped in the ball decomposed.

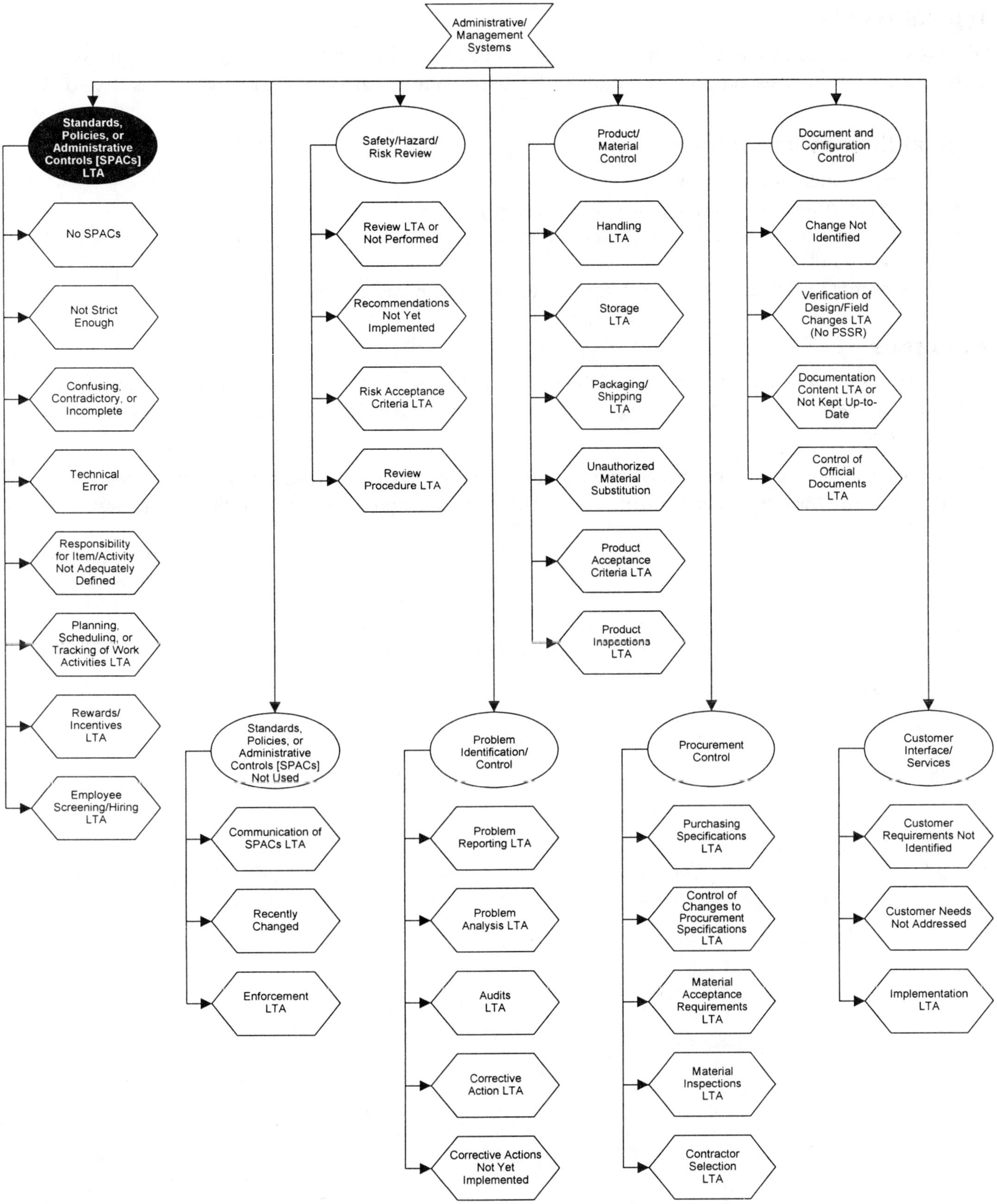
Administrative/ Management Systems
Standards, Policies, or Administrative Controls [SPACs] LTA
No SPACs
Not Strict Enough
Confusing, Contradictory, or Incomplete
Technical Error
Responsibility for Item/Activity Not Adequately Defined
Planning, Scheduling, or Tracking of Work Activities LTA
Rewards/ Incentives LTA
Employee Screening/Hiring LTA
Standards, Policies, or Administrative Controls [SPACs] Not Used
Communication of SPACs LTA
Recently Changed
Enforcement LTA
Safety/Hazard/ Risk Review
Review LTA or Not Performed
Recommendations Not Yet Implemented
Risk Acceptance Criteria LTA
Review Procedure LTA
Problem Identification/ Control
Problem Reporting LTA
Problem Analysis LTA
Audits LTA
Corrective Action LTA
Corrective Actions Not Yet Implemented
Product/ Material Control
Handling LTA
Storage LTA
Packaging/ Shipping LTA
Unauthorized Material Substitution
Product Acceptance Criteria LTA
Product Inspections LTA
Procurement Control
Purchasing Specifications LTA
Control of Changes to Procurement Specifications LTA
Material Acceptance Requirements LTA
Material Inspections LTA
Contractor Selection LTA
Document and Configuration Control
Change Not Identified
Verification of Design/Field Changes LTA (No PSSR)
Documentation Content LTA or Not Kept Up-to-Date
Control of Official Documents LTA
Customer Interface/ Services
Customer Requirements Not Identified
Customer Needs Not Addressed
Implementation LTA

Typical Issues

Was the error caused by the lack or inadequacy of SPACs? Were the SPACs inaccurate, confusing, incomplete, unclear, ambiguous, not strict enough, or otherwise inadequate? Were the wrong actions rewarded?

Typical Recommendations

- Provide written documentation of SPACs
- When errors are found, modify SPACs accordingly
- Ensure that policies regarding production, material control, procurement, security, etc., do not contradict safety policies
- Ensure that the rewards and incentives are consistent with facility objectives

Examples

An operator was unable to read at the level needed to understand facility procedures because employee screening standards were not high enough. As a result, he made a serious mistake in operating a key piece of equipment.

The surveillance testing for the fire protection system had not been conducted for the past 2 years. The maintenance and operations departments both thought the other group was responsible for the test.

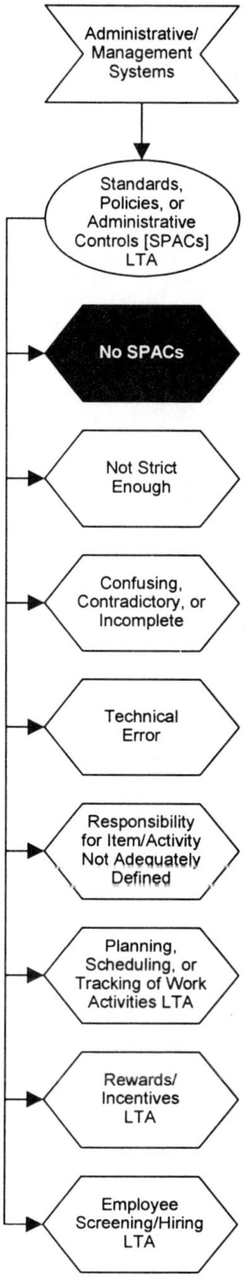

Typical Issues

Did a SPAC exist to control the particular type of work or situation involved in the incident? Was the work or situation significant or involved enough to warrant some type of SPAC to ensure adequate job quality and work control?

Typical Recommendations

- Compile a list of SPACs mandated by regulatory requirements (OSHA, EPA, etc.) and compare it to a current list of existing SPACs. Develop any missing SPACs
- Provide written documentation of SPACs
- Define, document, and communicate missing SPACs

Example

A maintenance worker was exposed to a pressurized release of a process material. The line from which the material was released had not been depressurized and cleared before maintenance work began. The plant did not have a safe work practice/permit for opening process equipment (i.e., "line breaking").

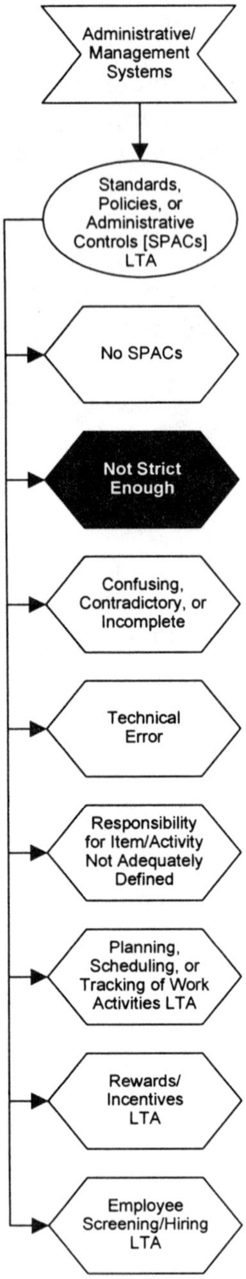

Typical Issues

Were the existing SPACs strict enough to provide adequate job quality or work control? Did vagueness allow violation of the intent, if not the letter, of the SPACs?

Typical Recommendations

- Improve the level of detail of SPACs
- Improve the description of accountabilities in SPACs (for resolving ambiguities)

Examples

A safety limit was violated during operation of a process because an alarm indicating a high temperature was bypassed. The first-line supervisor thought the alarm was false and bypassed it. The SPACs were not strict enough, because they allowed the supervisor to bypass an alarm without getting any review or approvals from management and technical support.

Operators were supposed to check the operation of pumps by taking vibration readings and checking temperatures. Instead, the operators just performed a visual check of the pumps. The SPACs did not specify how the operators were to check the pumps, and the SPACs did not require supervisors to periodically check on how the operators were performing their tasks.

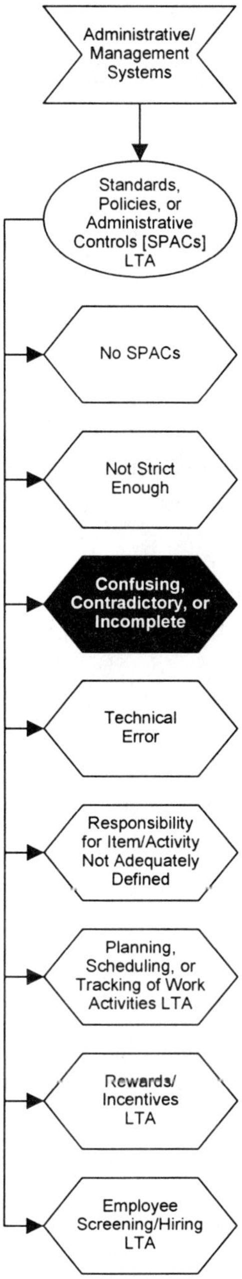

Typical Issues

Were the SPACs confusing, hard to understand or interpret, or ambiguous? Were the SPACs incomplete or not specific enough? Did contradictory requirements exist? Were some requirements violated or disregarded in order to follow others? Was a SPAC not followed because no practical way of implementing the SPAC existed? Would implementation have hindered production?

Typical Recommendations

- Solicit comments and recommendations from operations/maintenance personnel regarding ambiguous or unclear language in the SPACs. Resolve comments
- Ensure that policies regarding production, material control, procurement, security, etc., never contradict a safety policy
- Communicate to operators that safety should be given top priority
- Ensure that SPACs reflect management's decision to make safety top priority
- SPACs that require specific authorization signatures should state alternate sources of authorization in the event the primary authorizers are not available
- Provide the necessary tools/equipment features to allow/encourage personnel to follow the SPACs

Examples

A key piece of equipment in a process safety system failed. The policy stated that the required maintenance and inspections were to be performed annually. Because of the difficulty of scheduling the work with production and the amount of work involved, the maintenance/inspection cycle had gradually slipped to 18 months. The policy did not state a maximum period of 12 months.

A plant policy indicated that all "fatigue-related failures" be reported to the Equipment Reliability group. However, the maintenance organization had no guidance on what sort of failures were "fatigue-related." In addition, a recent reorganization resulted in the elimination of the Equipment Reliability group, and their previous functions were split among four other groups.

A release of a flammable liquid was larger than expected, overflowing the tank's dike. Administrative controls on the maximum intended inventories for the tanks in the dike were violated because of an anticipated shortage of the material from the supplier. Production decided to "stock up" on the material to prevent production outages.

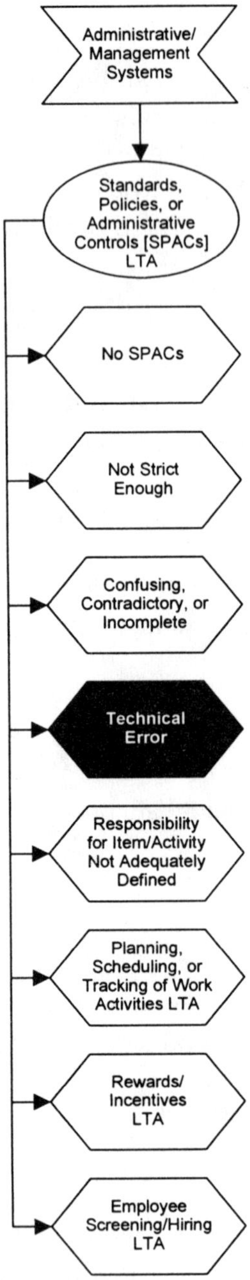

Typical Issues

Did technical errors or incorrect facts exist in the SPACs? Did the SPACs fail to consider all possible scenarios or conditions?

Typical Recommendations

- Include SPACs in the scope/charter of hazard review teams
- When errors are found, modify SPACs accordingly

Examples

A fire occurred when hot work near a process unit ignited vapors leaking from a nearby flange. The hot work policy for the plant erroneously indicated that a hot work permit was not necessary because the work was not specifically on equipment associated with hazardous materials.

Drawings were not updated following modifications. The SPAC controlling the design process was incorrect. Drawings were not sent to the correct individual in the drawing control department.

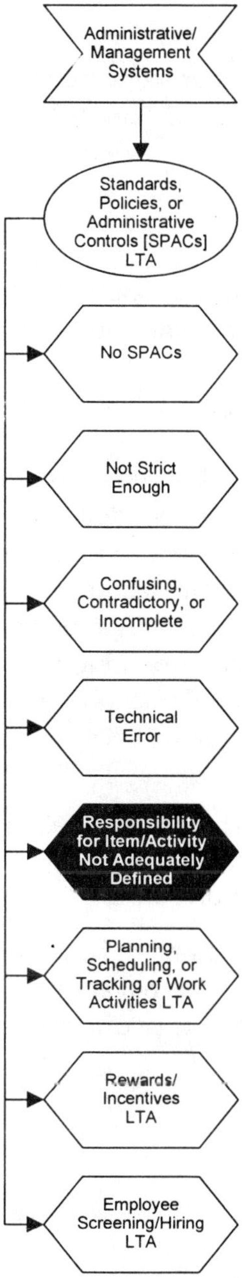

Typical Issues

Did the SPACs define the organization or group responsible for the item? Did confusion exist over who was responsible for the activity? Did an activity exist for which no one took responsibility?

Typical Recommendations

- Assign responsibility for items/activities by including specific job titles in SPACs
- Include accountability in job performance criteria (for job performance appraisals)

Example

A technical limit for the length of time allowed between air flow checks on a stack exhaust system was violated. The operations department considered the checks to be maintenance items. The maintenance department considered the checks to be an operations item. Responsibility for the checks was not defined.

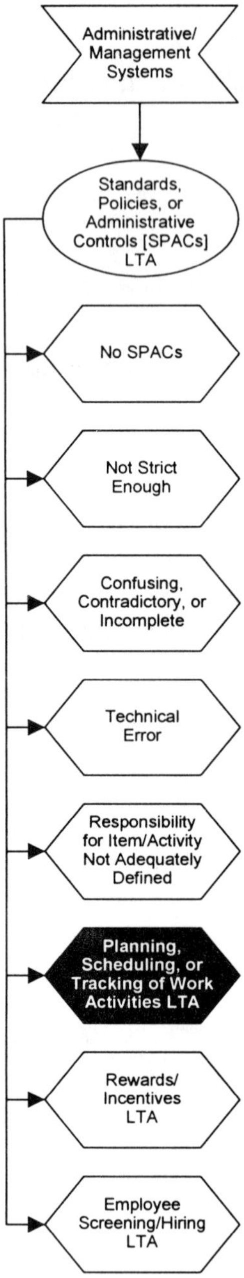

Typical Issues

Was the work scheduling system adequate? Was the work properly planned? Was the work schedule used for implementing work? Was the work scheduled based on safety and reliability impact?

Typical Recommendations

- Update the tracking system daily, weekly, or monthly, as appropriate, by adding new action items and/or documenting the current status of all action items
- Conduct periodic, unannounced audits to verify that those action items documented as "complete" are actually complete
- Limit access to the tracking/scheduling system to authorized personnel (e.g., use a password for an electronic system, lock system documentation in a filing cabinet and distribute keys only to authorized personnel)
- Prioritize action items and assign realistic dates for completion
- Develop and use "indicators" to help detect problems in ongoing use of management systems (e.g., how long does it take to respond to a request for change to a standard operating procedure?)

Examples

A tank overflowed during filling because the automatic shutoff valve failed to close. An earlier inspection found that the level switch for the valve was defective, but the equipment deficiency had not been resolved.

A scheduling system was developed by the maintenance planner; however, because there were too many panic repairs, the schedule was never followed. No one actually used the scheduling system to determine the priorities of the work that was performed.

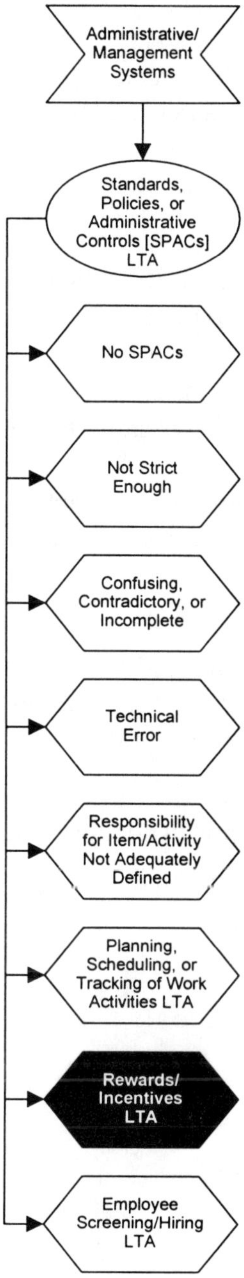

Typical Issues

Were workers rewarded for improper performance? Were incentives consistent with the goals of the company and facility? Did the reward system encourage workers to take short cuts or waste resources?

Typical Recommendations

- Develop rewards that are consistent with company goals and objectives
- Ensure that metrics and other measurements for performance are consistent with facility goals and objectives

Examples

Performance of customer service representatives was measured by the number of calls they handled each day. As a result, they tried to diagnose the problem as quickly as possible and provide a recommended solution. Because the representatives were trying to diagnose the problem quickly, they often misdiagnosed the problem. About 40% of the phone calls were repeat calls from customers whose problems were misdiagnosed the first time.

One of the metrics for the maintenance organization was the percentage of utilization for a certain lathe. This was measured by the percentage of time the lathe was operating. As a result, the operators turned on the lathe in the morning when they came in and let it run until they went home. They never used the lathe for work because it would decrease the amount of time the machine ran.

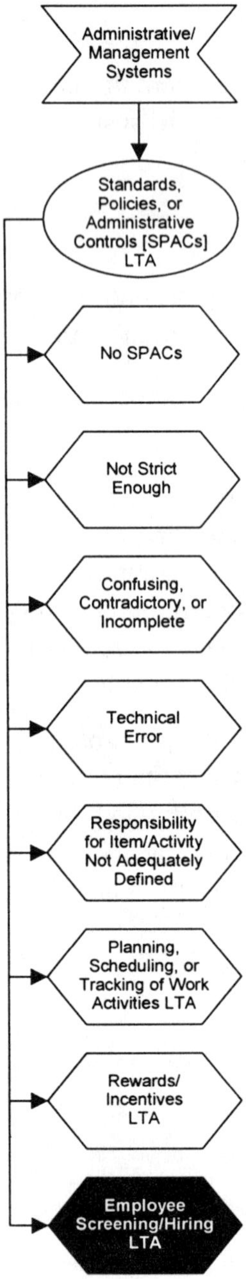

Typical Issues

Did an effective employee screening program exist? Did it correctly identify requirements for particular jobs? Did it screen employees against those requirements?

Typical Recommendations

- Assess critical personal capability requirements for each job position
- Communicate all required job tasks to potential employees before extending employment opportunities
- Ask job interviewees if they can perform job-related tasks
- Consider requiring the passing of a physical exam/drug-screen test as a contingency of employment
- Have prospective employees perform a test that simulates the actual work as closely as possible to determine if they can perform the work

Examples

An operator made a mistake operating a process on a color-coded distributed control system because he was color blind. Although a screening program existed for the job, it did not specify the ability to differentiate colors as a requirement.

A maintenance technician made an error in repairing a mill. The technician could not read the procedure that he was supposed to use.

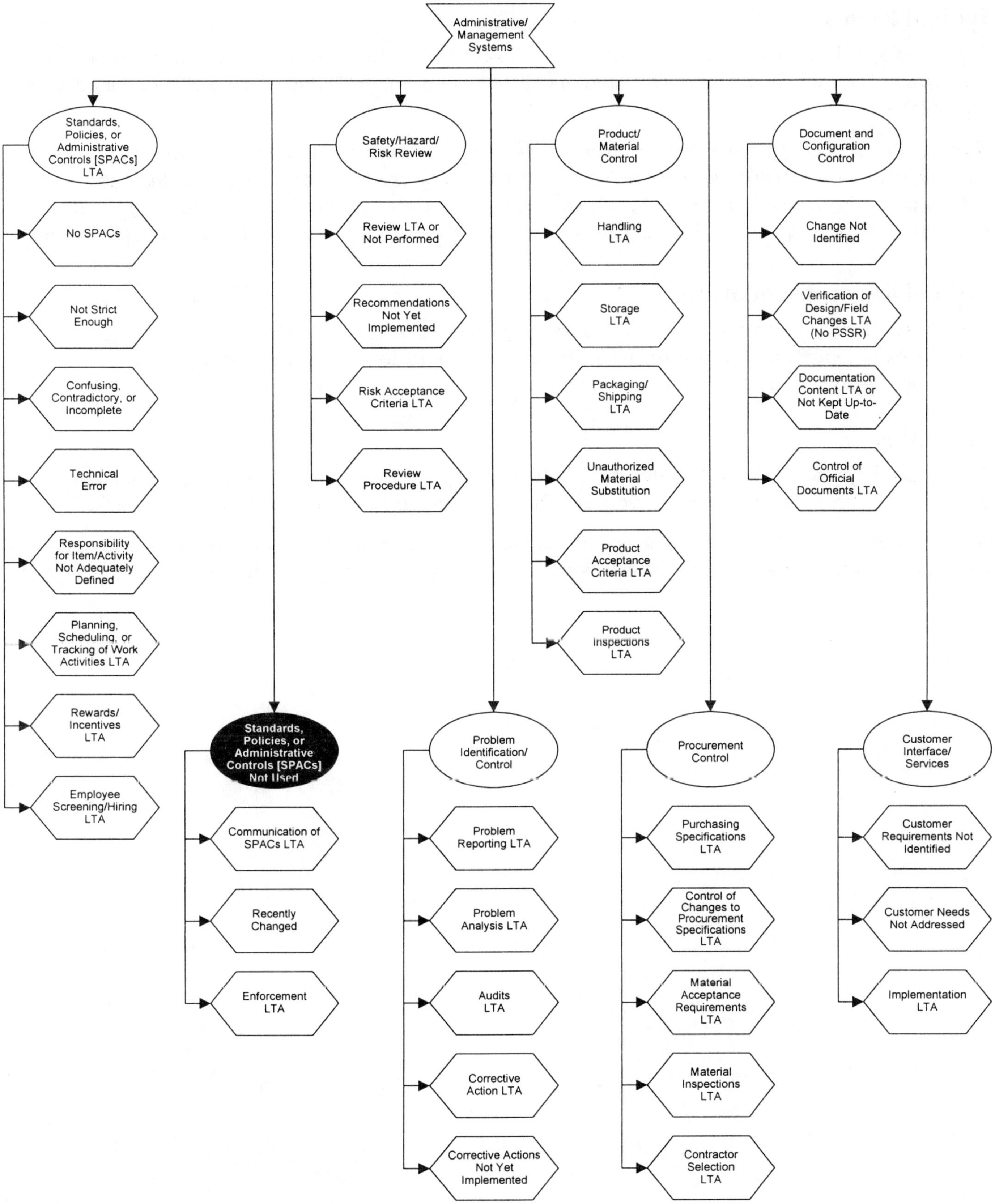
Administrative/ Management Systems
Standards, Policies, or Administrative Controls [SPACs] LTA
No SPACs
Not Strict Enough
Confusing, Contradictory, or Incomplete
Technical Error
Responsibility for Item/Activity Not Adequately Defined
Planning, Scheduling, or Tracking of Work Activities LTA
Rewards/ Incentives LTA
Employee Screening/Hiring LTA
Standards, Policies, or Administrative Controls [SPACs] Not Used
Communication of SPACs LTA
Recently Changed
Enforcement LTA
Safety/Hazard/ Risk Review
Review LTA or Not Performed
Recommendations Not Yet Implemented
Risk Acceptance Criteria LTA
Review Procedure LTA
Problem Identification/ Control
Problem Reporting LTA
Problem Analysis LTA
Audits LTA
Corrective Action LTA
Corrective Actions Not Yet Implemented
Product/ Material Control
Handling LTA
Storage LTA
Packaging/ Shipping LTA
Unauthorized Material Substitution
Product Acceptance Criteria LTA
Product Inspections LTA
Procurement Control
Purchasing Specifications LTA
Control of Changes to Procurement Specifications LTA
Material Acceptance Requirements LTA
Material Inspections LTA
Contractor Selection LTA
Document and Configuration Control
Change Not Identified
Verification of Design/Field Changes LTA (No PSSR)
Documentation Content LTA or Not Kept Up-to-Date
Control of Official Documents LTA
Customer Interface/ Services
Customer Requirements Not Identified
Customer Needs Not Addressed
Implementation LTA

Typical Issues

Were SPACs or directives not used, adhered to, or followed? Was communication or enforcement of SPACs inadequate? Were the SPACs recently revised or difficult to implement? Did the SPACs provide for adequate accountability?

Note: *SPACs provide guidance on how an activity should be accomplished, whereas procedures provide a detailed, step-by-step method for performing a specific task. For example, there are SPACs that describe the policies governing scheduling of workers. There is also a procedure that provides a detailed, step-by-step process for performing the task, including the forms to complete and data to enter in the computer system.*

Typical Recommendations

- Ensure that all levels of affected employees are aware of SPACs changes
- Take appropriate actions concerning those employees who do not use the SPACs
- Apply lessons learned from one unit to other units

Examples

A mechanic bypassed an important step in calibrating a key safety instrument because he did not take a printout of the procedure with him, as required. This was found to be an accepted practice in the facility.

A requirement was in place to have the operators check instruments in the field once per shift. The operators never performed the checks. Supervision was aware of the situation and never enforced the requirement.

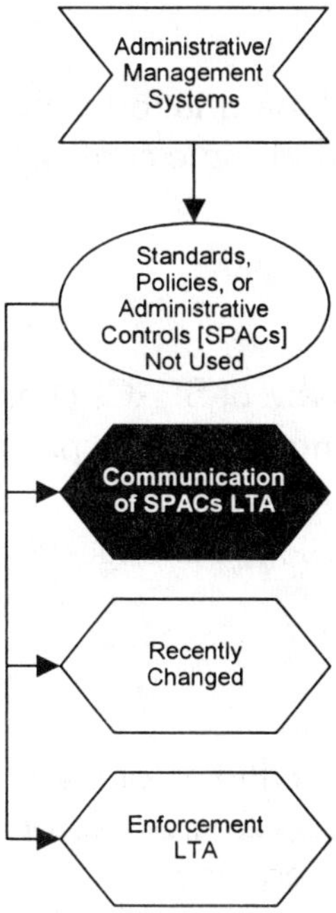

Typical Issue

Were standards, directives, or policies not communicated from management down through the organization?

Typical Recommendations

- Include SPACs content in initial and refresher formal training; determine employee's understanding
- Periodically stress the importance of using SPACs during shift change meetings, safety meetings, etc.
- Ensure that SPACs documentation is readily available to all affected employees at all times for reference purposes

Example

During an extended facility outage, routine surveillances of process alarm panels were not performed. As a result, a chemical leak went undetected for 2 days. Facility management had not communicated to first-line supervisors that normal surveillance procedures remained in effect during the outage.

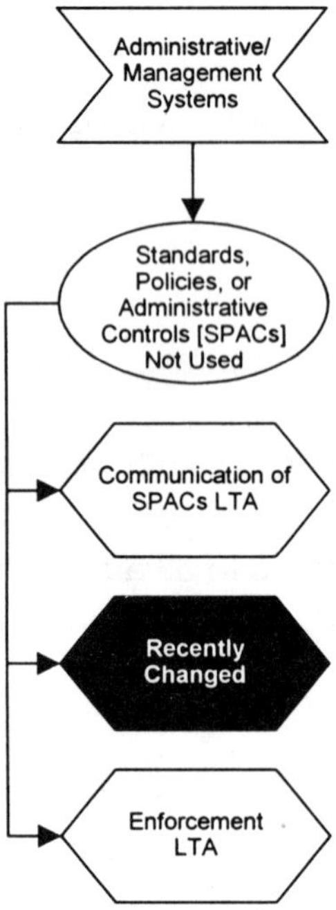

Typical Issues

Had standards or directives been recently changed? Did information concerning changes fail to reach all levels of the organization? Had some confusion been created by the changes?

Typical Recommendations

- Ensure that all levels of affected employees are aware of SPACs changes
- Verify that employees fully understand recent changes before expecting them to implement the changes
- Ensure that there is a process for communicating SPACs changes to the individuals who need to know about the changes

Examples

A new policy on calibration of flow indicators was provided to all of the maintenance department supervisors, but the mechanics were not told of the change. As a result, the policy was not implemented, as required.

A new policy was put in place to require personnel to enter the time charged against each work order into a computer system. No one was told of the requirement or taught how to enter the information in the computer.

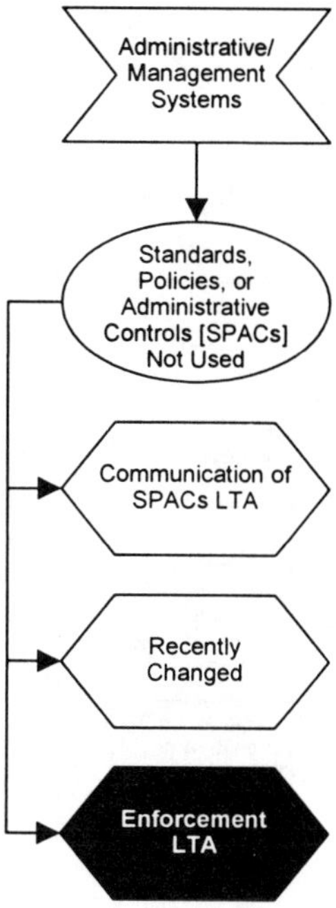

Typical Issues

In the past, has enforcement of the SPAC been lax? Have failures to follow the SPAC in the past gone uncorrected or unpunished? Has noncompliance been accepted by management and supervision?

Note: *Coding under* Rewards/Incentives LTA (SPACs/LTA), *or* Improper Performance Not Corrected (Immediate Supervision, Supervision During Work) *may be appropriate.*

Typical Recommendations

- Management should set an example by always following the letter of the SPACs
- Employees who do not use the SPACs should be corrected and/or punished
- Discipline needs to be fair, impartial, pre-stated, sure, and swift
- Enforcement needs to be consistent

Examples

A mechanic made a mistake installing a piece of equipment. He did not refer to a procedure when performing the test. Although policy is to always refer to the procedure, the policy had not been enforced. Mechanics often did not take procedures to the work site, and their supervisors were aware of this.

Operators were supposed to log local tank levels every 2 hours. However, they would typically take the readings only at the beginning of the shift. They used these readings to fill in the readings for the remainder of the shift. No one ever took issue with this practice until after an accident occurred.

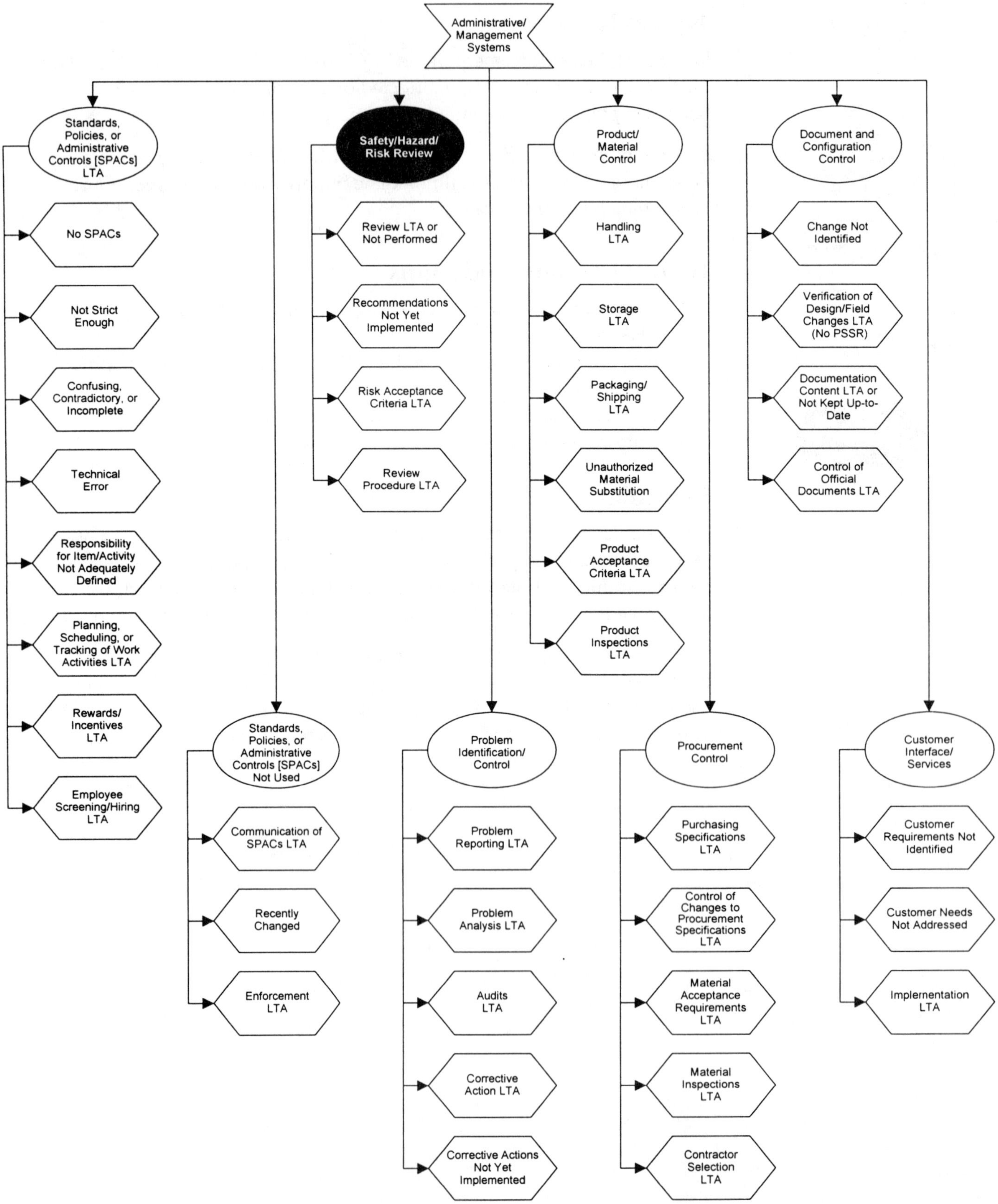
Administrative/ Management Systems
Standards, Policies, or Administrative Controls [SPACs] LTA
No SPACs
Not Strict Enough
Confusing, Contradictory, or Incomplete
Technical Error
Responsibility for Item/Activity Not Adequately Defined
Planning, Scheduling, or Tracking of Work Activities LTA
Rewards/ Incentives LTA
Employee Screening/Hiring LTA
Safety/Hazard/ Risk Review
Review LTA or Not Performed
Recommendations Not Yet Implemented
Risk Acceptance Criteria LTA
Review Procedure LTA
Product/ Material Control
Handling LTA
Storage LTA
Packaging/ Shipping LTA
Unauthorized Material Substitution
Product Acceptance Criteria LTA
Product Inspections LTA
Document and Configuration Control
Change Not Identified
Verification of Design/Field Changes LTA (No PSSR)
Documentation Content LTA or Not Kept Up-to-Date
Control of Official Documents LTA
Standards, Policies, or Administrative Controls [SPACs] Not Used
Communication of SPACs LTA
Recently Changed
Enforcement LTA
Problem Identification/ Control
Problem Reporting LTA
Problem Analysis LTA
Audits LTA
Corrective Action LTA
Corrective Actions Not Yet Implemented
Procurement Control
Purchasing Specifications LTA
Control of Changes to Procurement Specifications LTA
Material Acceptance Requirements LTA
Material Inspections LTA
Contractor Selection LTA
Customer Interface/ Services
Customer Requirements Not Identified
Customer Needs Not Addressed
Implementation LTA

Typical Issues

Was the error caused by an inadequate hazard review of the system? Was a risk assessment of the system performed? Have the safety and reliability hazards been identified?

Typical Recommendations

- Ensure that all newly installed and/or modified equipment is included in a hazard review prior to startup
- Track and document the final resolution for all recommendations
- Ensure that personnel, equipment, and environmental losses are all addressed in the review

Examples

A wastewater tank was overpressurized and failed. During the tie-in of a new line, the review team recommended the installation of a larger overflow line to handle the largest possible flow into the tank. The results of the review were not incorporated into the installation package. The new line was tied into the tank without a new overflow line installed.

A scenario for rapid overpressurization of an atmospheric decanter system was not considered prior to startup of a process because the hazard review did not address procedural deviations during an allowable startup mode.

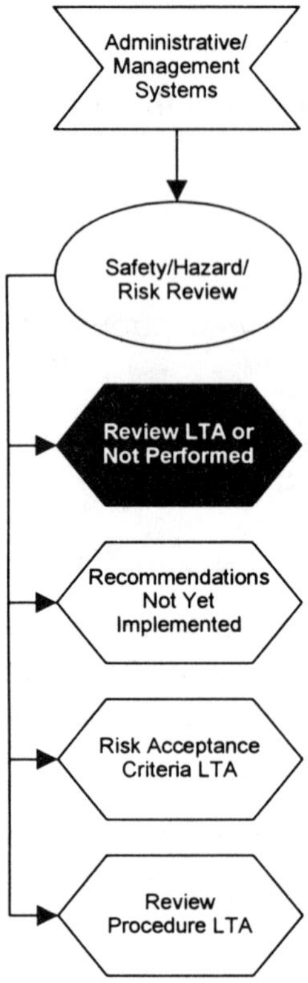

Typical Issues

Was the safety and hazard review complete? Did it consider all modes of operation/maintenance, and were other required hazard review issues considered? Was the review done according to all applicable orders, regulations, and guides? Was a safety/hazard/risk review performed?

Typical Recommendations

- Provide a safety/hazard/risk review procedure that complies with all applicable orders, regulations, and guides
- Ensure that the hazard review procedure is readily available to personnel who will conduct the review
- Periodically audit hazard review meetings and reports
- Establish minimum training criteria for hazard review leaders
- Ensure that all newly installed and/or modified equipment is included in a hazard review prior to startup
- Ensure that hazard review documentation is readily available to document the content of the review and to confirm that a review was performed

Examples

An explosion occurred in a waste tank because incompatible materials were mixed. The process hazards review had been performed, but it failed to consider all the possible sources of material that could be added to the tank.

An explosion occurred in a waste tank after a new stream had been tied into the tank. No safety review had been performed prior to tying in the stream to determine if incompatible materials would be in the waste tank after the tie in.

No analysis had been performed to determine the operational risks associated with a new conveyor system.

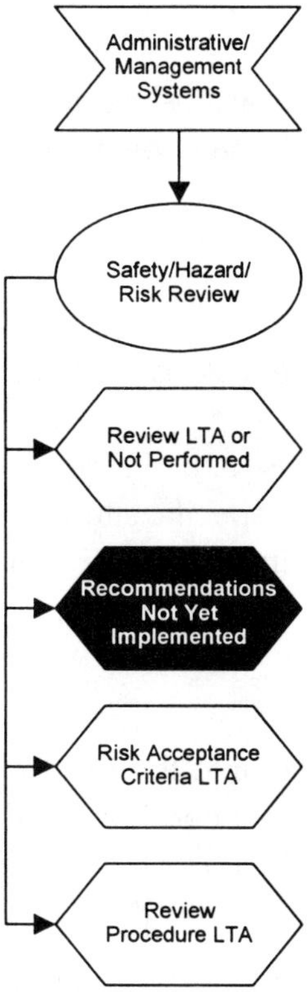

Typical Issue

Have the recommendations from the safety/hazard/risk review been implemented?

Typical Recommendations

- Ensure that all hazard review recommendations are documented and reviewed by management personnel
- Management should address all hazard review recommendations and document the manner in which the recommendation will be resolved (i.e., assign a responsible party for completion or reject the recommendation with documented reason for doing so)
- Communicate hazard review recommendations to all affected parties
- Document the final resolution or implementation of each recommendation
- Publish periodic reports of resolution status for management
- Ensure that implementation of the recommendations is assigned to a specific group or individual

Examples

A release of hazardous material through a rupture disk was discharged to the diked area of the process. The hazard review had recommended installing a catch tank, with a rain hood/cover, to receive any discharged material. The catch tank had not been installed because of scheduling conflicts with other construction in the area. The released material reacted violently with rain water in the diked area, producing a large quantity of toxic gas.

As a result of a facility risk/reliability analysis, recommendations were made to have a final inspection performed of unusual and partial shipments to ensure that they are correct. This recommendation had not been implemented yet.

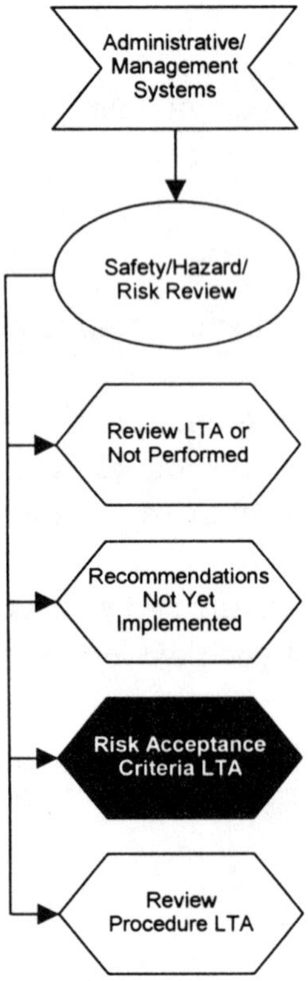

Typical Issues

Were the risk acceptance criteria used during the safety/hazard/risk review set inappropriately? Were risks deemed acceptable that should have been reduced?

Typical Recommendations

- Ensure that a diverse team (able to reasonably assess risk) is involved in the hazard review
- Develop more objective criteria for judging risk levels (e.g., a simplified risk scoring scheme or listing required safeguards for specific situations)
- Provide guidance to team members to help ensure that the reviews are conducted properly

Examples

An explosion occurred when the incorrect material was fed into the reactor. The supplier had mislabeled the material. The hazard review had identified this as a risk factor but concluded that the risks associated with not analyzing the incoming materials were acceptable.

Company criteria for multiple layers of safeguards allowed a large credit for relief valves. As a result, insufficient attention was given to reducing the frequency of relief valve actuations.

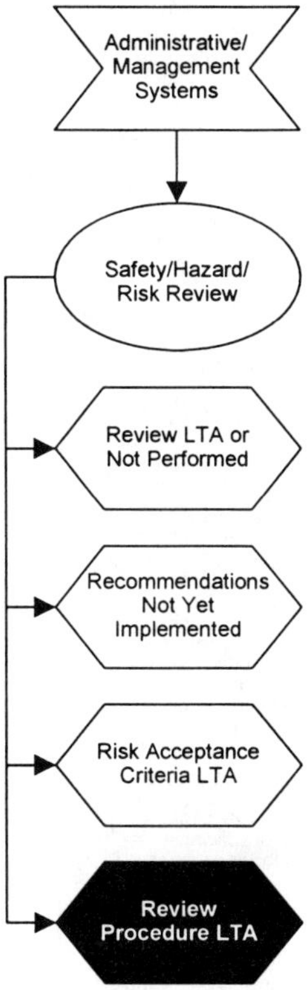

Typical Issues

Was the safety/hazard/risk review procedure less than adequate? Does it provide adequate guidance for the scope of the review? Are the resources needed to perform the review available? Are personnel trained in the use of the procedure?

Typical Recommendations

- Ensure that the hazard review technique is appropriate for the complexity of the process
- Ensure that all newly installed and/or modified equipment is included in a hazard review prior to startup
- Ensure that hazard reviews comply with all applicable orders, regulations, and guides (e.g., some provide specific checklists for the safety/hazard review)
- Ensure that the review procedure addresses scope of analyses and training of hazard analysis team leaders

Examples

A complex shutdown system failed to mitigate a process upset, resulting in a release of a hazardous material. The review procedure for the plant specified that a HAZOP be performed for all new/modified systems; however, the HAZOP procedure was not well-suited for analyzing this type of system (the FMEA technique would have been a better choice of technique).

A major spill violating an environmental permit occurred at a process that had recently undergone a hazard review. This type of spill, which had no safety consequences, was not addressed in the study because the review procedure did not require evaluation of environmental hazards.

A risk assessment was recently performed on a packaging operation. The risk assessment did not address supply problems because the review procedure did not require that issue to be considered. Later, a fire at a key supplier's facility led to a 4-week shutdown.

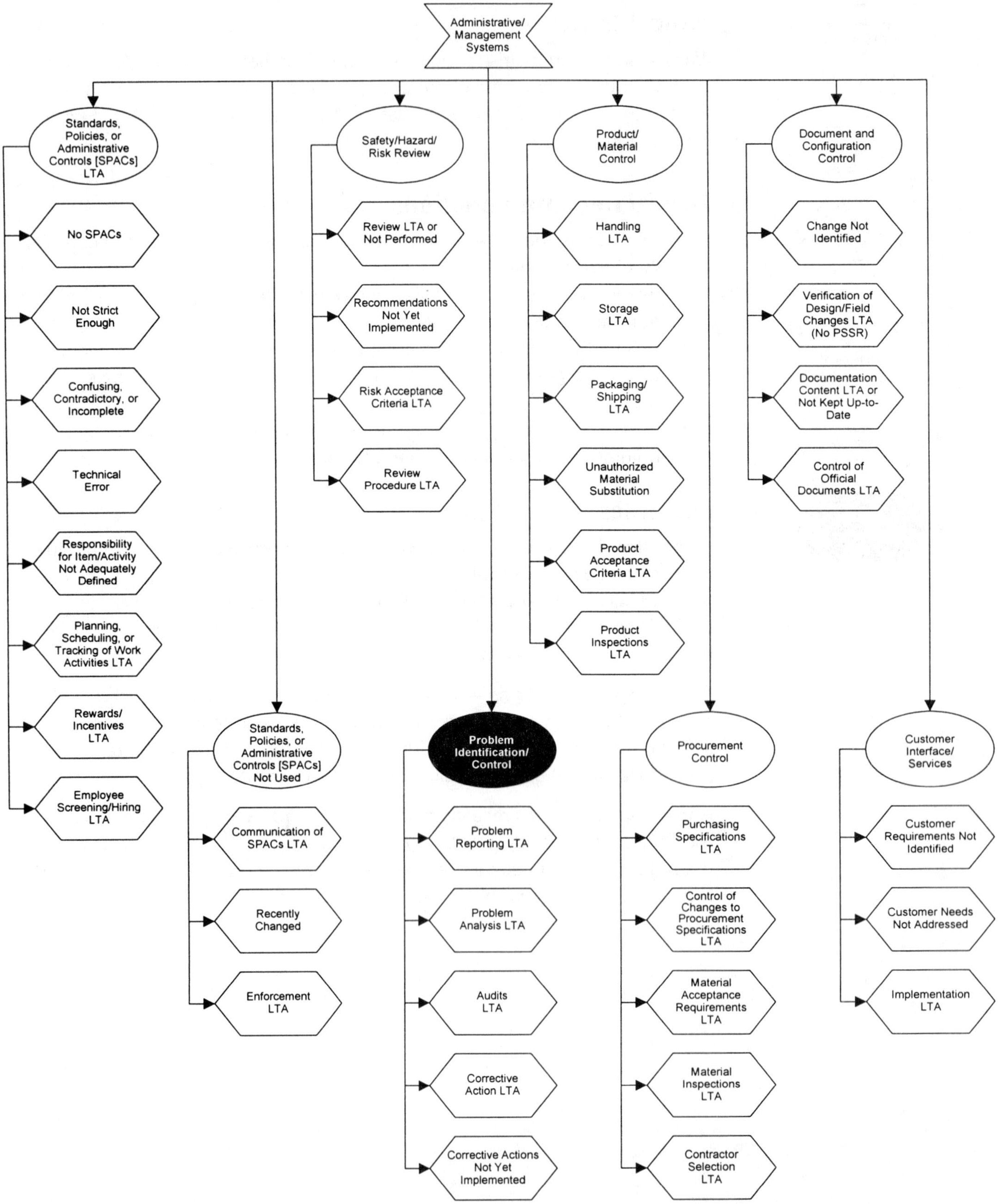

Administrative/ Management Systems
Standards, Policies, or Administrative Controls [SPACs] LTA
No SPACs
Not Strict Enough
Confusing, Contradictory, or Incomplete
Technical Error
Responsibility for Item/Activity Not Adequately Defined
Planning, Scheduling, or Tracking of Work Activities LTA
Rewards/ Incentives LTA
Employee Screening/Hiring LTA
Safety/Hazard/ Risk Review
Review LTA or Not Performed
Recommendations Not Yet Implemented
Risk Acceptance Criteria LTA
Review Procedure LTA
Product/ Material Control
Handling LTA
Storage LTA
Packaging/ Shipping LTA
Unauthorized Material Substitution
Product Acceptance Criteria LTA
Product Inspections LTA
Document and Configuration Control
Change Not Identified
Verification of Design/Field Changes LTA (No PSSR)
Documentation Content LTA or Not Kept Up-to-Date
Control of Official Documents LTA
Standards, Policies, or Administrative Controls [SPACs] Not Used
Communication of SPACs LTA
Recently Changed
Enforcement LTA
Problem Identification/ Control
Problem Reporting LTA
Problem Analysis LTA
Audits LTA
Corrective Action LTA
Corrective Actions Not Yet Implemented
Procurement Control
Purchasing Specifications LTA
Control of Changes to Procurement Specifications LTA
Material Acceptance Requirements LTA
Material Inspections LTA
Contractor Selection LTA
Customer Interface/ Services
Customer Requirements Not Identified
Customer Needs Not Addressed
Implementation LTA

Typical Issues

Was an event caused by failure to provide corrective action for known deficiencies or failure to implement recommended corrective actions before known deficiencies recur? Had the problem occurred before and never been reported? Did an audit fail to discover the problem? Did the corrective actions implemented fail to correct the problem?

Note: *If the problem/deficiency could/should have been identified, or was identified in a safety/hazard/risk review, then code the event in that portion of the Map and not here.*

Typical Recommendations

- Track implementation of corrective actions to ensure timely completion
- Consider implementing the same corrective actions for similar situations at this and other facilities
- Measure the effectiveness of corrective actions
- Periodically compare the results of audits with events that occur in the facility to ensure that audits are effective in identifying problems

Examples

A tank overflowed because an operator ignored an auditory alarm in the control system. The alarm, which sounded spuriously about every 15 minutes, had been broken for more than 6 months. The problem had been reported to the maintenance organization but had not been repaired.

A tank had overflowed when the operator started the wrong pump. None of the pump control switches were labeled. A corrective action from this event was to install labels on the pump switches. Prior to installation of the labels, another pump was damaged when the operator started the wrong pump. The switches for these pumps were not labeled either.

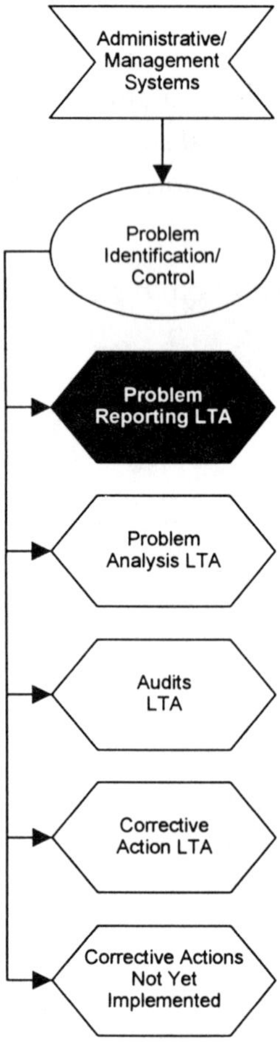

Typical Issues

Are personnel reporting events that have significant impacts on health, safety, or reliability? Are personnel aware of the types of events that should be reported? Do they know how to report the events? Are employees punished for reporting problems?

Note: *Coding under* Rewards/Incentives LTA *may also be appropriate.*

Typical Recommendations

- Develop event-reporting guidelines
- Provide training to personnel on the types of events that should be reported. Make these examples as process specific as possible
- Ensure that the event-reporting process is as simple as possible

Examples

An engineer noted oil dripping from a pump seal. The process for reporting and documenting the problem required a lot of forms to be filled out. The engineer did not want to take the time to complete the forms. As a result, he did not report the problem.

An operator reported a problem with the drying oven he was using. The temperature control system had malfunctioned and a batch of product had been damaged. Company policy required individuals who reported problems to help personnel correct the situation. As a result, the operator was required to work overtime to assist with the repairs and he missed the college basketball championship game on television. The next time the operator discovered a problem near the end of his shift, he did not report it because he did not want to stay over past his shift.

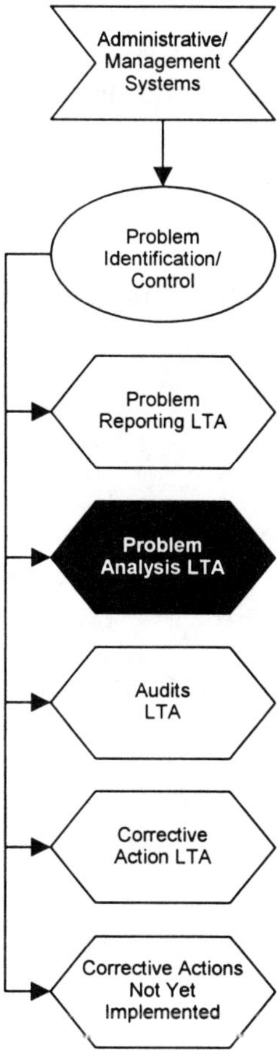

Typical Issues

Was the problem misdiagnosed? Were knowledgeable personnel involved in the problem analysis? Was proper emphasis placed on problem diagnosis?

Typical Recommendations

- Develop generic methods for problem analysis such as the 5 Whys technique, fault tree analysis, and/or causal factor charting
- Train all personnel to some level of troubleshooting. Provide appropriate experts to assist analysis teams
- Have the results of the analysis reviewed by someone outside the organization

Examples

An accident occurred in a reactor vessel. The incident investigation team thought the explosion was caused by a lack of grounding on the tank. After a second event, it was determined that the wrong materials were being fed into the tank and that this had triggered the explosion.

A root cause analysis team determined that spurious shutdowns of a mixing line were caused by operator errors. Subsequent shutdowns indicated that electronic spikes were causing pressure spikes that caused a safety system to actuate and shut down the line.

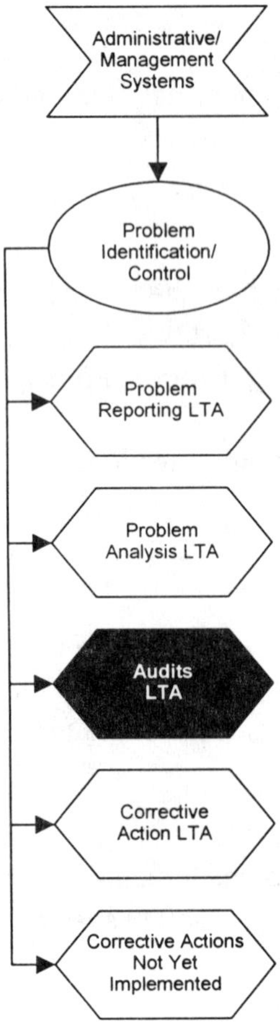

Typical Issues

Do audits find problems before they cause safety, reliability, or quality problems? Are audits performed at regular intervals?

Typical Recommendations

- Ensure that periodic audits of systems important to safety, reliability, and quality are developed
- Ensure that audits are periodically implemented

Examples

An audit had been developed to ensure that personal protective equipment (hard hats, safety goggles, etc.) was being worn by plant personnel. However, the audit was only conducted once.

No audits had been developed to determine if quality assurance inspections of final products were being implemented effectively.

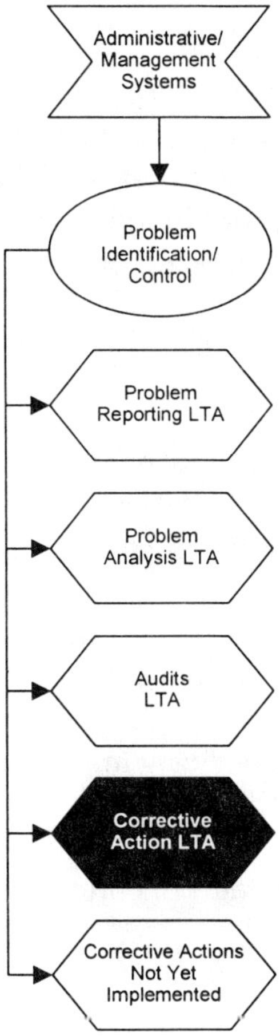

Typical Issues

Were implemented corrective actions unsuccessful in preventing recurrence? Should other corrective actions have been identified? Were corrective actions focused on correcting the root causes of the problem?

Typical Recommendations

- Involve a multidisciplinary team in identifying corrective actions to ensure that the problem has been fully analyzed
- Refer design/development of corrective actions to specialists when teams have difficulty identifying practical solutions
- Develop measures to determine the effectiveness of corrective actions
- Trend event causes and root causes to determine if corrective actions are effective in preventing recurrence

Examples

A problem with operators bypassing alarms had been identified. The corrective action was to administratively control alarm bypasses. After a couple of years, the administrative control requirements were being ignored. Physical changes to equipment may have been more successful in preventing bypassing of alarms.

The procedure development process was modified to ensure that precautions and warnings were placed in procedures where appropriate. However, an audit of procedures performed a year later identified hundreds of procedures that did not have the proper precautions and warnings.

An operator was fired for poor performance. The operator had produced a number of bad batches. An experienced operator was moved into this position and also produced a number of bad batches. When the system was analyzed, it was determined that the control system was poorly designed and could not be easily controlled.

A gear tooth failure destroyed the gear train of a printing press. Only those gears with visibly damaged teeth were replaced. The press failed again about 6 months later when another gear tooth, overstressed but not visibly deformed by the first incident, failed.

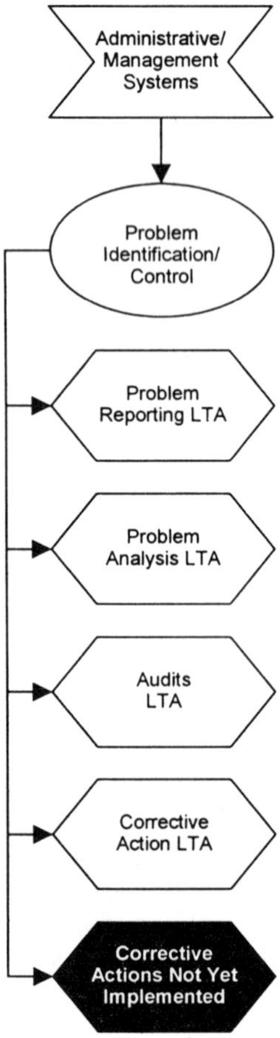

Typical Issue

Was a recommended corrective action for a known deficiency not implemented (because of delays in funding, delays in project design, normal length of implementation cycle, tracking deficiencies, etc.) before recurrence of the deficiency? Are corrective actions assigned to specific groups or individuals for implementation? Does management monitor the implementation of corrective actions?

Typical Recommendations

- If a system is deficient and requires corrective actions that cannot be implemented immediately, interim measures should be taken (implementing a temporary operating procedure, process parameter changes, shutting equipment down, etc.)
- Corrective actions affecting safety should not be delayed because of lack of funding, delays in project design, or normal length of the implementation cycle
- The cost of implementing corrective actions with significant impacts on reliability and quality should be balanced against the anticipated savings from implementation
- Ensure that management periodically reviews the status of corrective actions
- Reward personnel for completing corrective actions

Examples

A tank collapsed under vacuum. An earlier hazard analysis recommended vacuum breakers for this type of tank, but these devices had not yet been installed.

A root cause analysis of a quality problem recommended that special orders be packaged in different colored barrels to highlight the need for special handling. Since this recommendation was made, 16 more instances of mistakes with special orders occurred. The recommendation had never been implemented.

An incident investigation recommended that small drain holes be drilled in the discharge line of all fire monitors to prevent accumulation of water that could freeze and plug the monitor. This recommendation had not been implemented before another fire occurred, and two of the three monitors failed because they were plugged with ice.

An audit recommended shape coding of certain controls on the control panel to avoid selection errors. This recommendation had not been implemented when another batch of product was ruined as a result of an operator switch selection error.

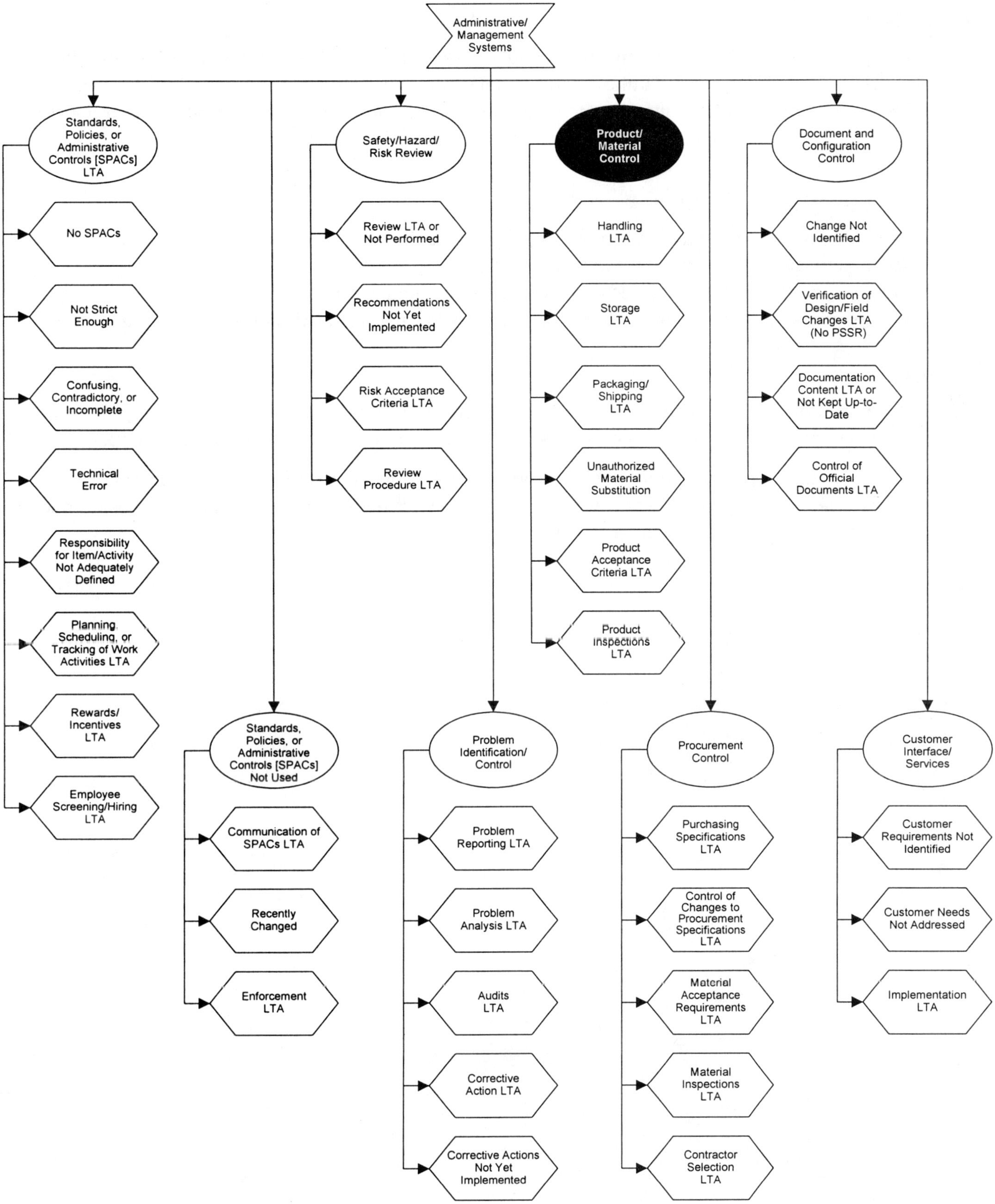
Administrative/ Management Systems
Standards, Policies, or Administrative Controls [SPACs] LTA
No SPACs
Not Strict Enough
Confusing, Contradictory, or Incomplete
Technical Error
Responsibility for Item/Activity Not Adequately Defined
Planning, Scheduling, or Tracking of Work Activities LTA
Rewards/ Incentives LTA
Employee Screening/Hiring LTA
Safety/Hazard/ Risk Review
Review LTA or Not Performed
Recommendations Not Yet Implemented
Risk Acceptance Criteria LTA
Review Procedure LTA
Product/ Material Control
Handling LTA
Storage LTA
Packaging/ Shipping LTA
Unauthorized Material Substitution
Product Acceptance Criteria LTA
Product Inspections LTA
Document and Configuration Control
Change Not Identified
Verification of Design/Field Changes LTA (No PSSR)
Documentation Content LTA or Not Kept Up-to-Date
Control of Official Documents LTA
Standards, Policies, or Administrative Controls [SPACs] Not Used
Communication of SPACs LTA
Recently Changed
Enforcement LTA
Problem Identification/ Control
Problem Reporting LTA
Problem Analysis LTA
Audits LTA
Corrective Action LTA
Corrective Actions Not Yet Implemented
Procurement Control
Purchasing Specifications LTA
Control of Changes to Procurement Specifications LTA
Material Acceptance Requirements LTA
Material Inspections LTA
Contractor Selection LTA
Customer Interface/ Services
Customer Requirements Not Identified
Customer Needs Not Addressed
Implementation LTA

Typical Issues

Was the problem caused by inadequate material handling, storage, packaging, or shipping? Was the shelf life for the material exceeded? Was an unauthorized material substitution made? Were spare parts inadequately stored? Was the problem caused by inadequate handling, storage, packaging, or shipping of finished products?

Typical Recommendations

- Ensure that materials are stored in a proper environment
- Inspect materials for damage upon arrival at the facility
- Provide proper packaging of finished products to avoid damage during shipping
- Provide proper environmental conditions for raw materials and finished products to ensure quality

Examples

As a result of improper labeling, a grease was placed into inventory on the wrong shelf in the supply room. Subsequently, a pump failed when this grease was used instead of the one specified for that pump.

Because of a snow storm, product could not be shipped on schedule. The warehouse was full of finished product so it was temporarily stored in narrow aisles in the process area. Some of the product was damaged when an operator ran into the skids with a forklift.

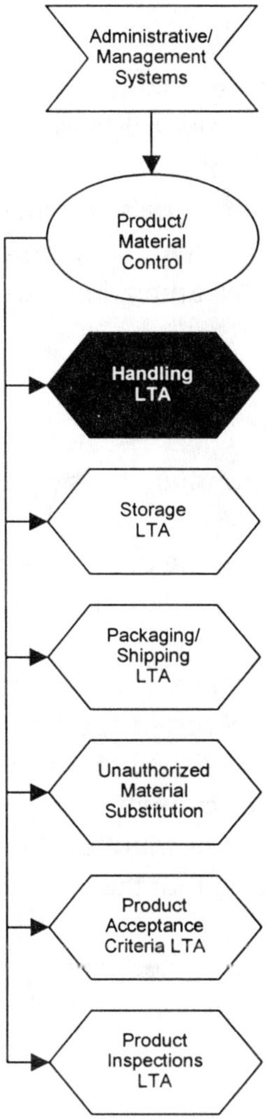

Typical Issues

Was material/equipment/product damaged during handling? Were items "mixed up" during handling? Was the equipment used for moving materials appropriate for the items?

Typical Recommendations

- Consider the size, weight, and hazards associated with transporting materials, and choose a mode of transport that is appropriate
- Consider tagging equipment before transporting it to the field to ensure that similar items are not inadvertently switched

Examples

The wrong pump was installed in a line. The mechanics were installing several pumps and had them all on a cart. They were "mixed up" and installed in the wrong locations.

Machined plates were placed in a cart for transport from one station to the next. Recently, the cart had been repaired with screws that were too long. As a result, some of the plates were scratched as they were placed on or removed from the cart.

A conveyor that was used to move brass fixtures to a packaging location moved continuously. This caused the fixtures to rotate for long periods of time while they were bunched up near the end of the conveyor. As a result, the finish on some of the fixtures was marred.

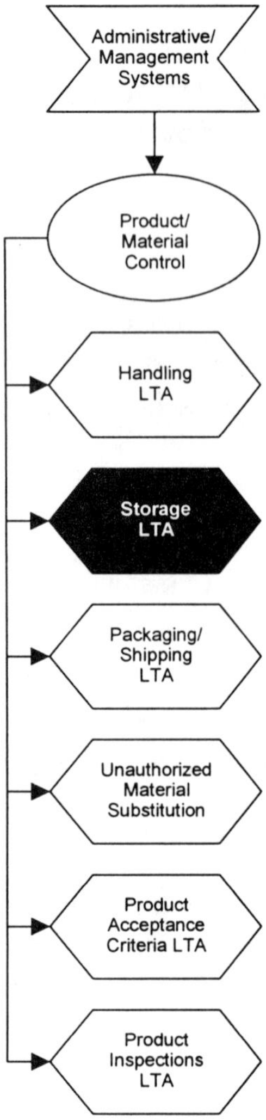

Typical Issues

Was material stored improperly? Was it damaged in storage? Did it have weather damage? Was it stored in an environment (heat, cold, acid fumes, etc.) that damaged it? Was product properly stored? Were material/equipment/parts issued after their shelf life was exceeded? Did materials continue to be used after the shelf life was exceeded? Were spare parts and equipment stored properly? Was adequate preventive maintenance (cleaning, lubrications, etc.) performed on spares?

Typical Recommendations

- Ensure that materials that require a controlled environment for storage are not exposed to the weather
- Before stacking materials in a warehouse, ensure that the contents and the packaging are compatible with this storage configuration and will not be damaged
- Promptly correct problems affecting storage in controlled environments (failures of heating/cooling systems, humidity control systems, etc.)
- Ensure that the proper environment is provided for finished product
- For materials with a shelf life, develop a system to document the material's shelf life, date of manufacture, and date of distribution
- Assign stores employees the responsibility of ensuring that the shelf life has not been exceeded
- Ensure that spare parts are not exposed to adverse weather conditions
- Promptly correct problems in equipment storage conditions or environmental controls in warehouses

Examples

An absorption column installed to remove contaminants from solvent did not operate as designed. Investigation revealed that the absorbent material used to pack the column had been stored outside and uncovered. The damaged material reduced the efficiency of the column.

The air conditioning system in the finished product storage area at a glue factory was inoperable for about a week during the summer. The warehouse reached temperatures of over 120 °F. Some of the glues were damaged from the excessive heat.

Rubber tubing used in the cooling system of portable generators cracked and failed. The shelf life of the rubber tubing installed had been exceeded and the tubing had become brittle.

A pump failed shortly after installation, which was much earlier than anticipated given the life expectancy of the pump. Investigation revealed that the pump had been stored in spare parts for a long time. During the storage, no preventive maintenance, such as cleaning and lubrication, had been performed as specified in the manufacturer's instructions for storage.

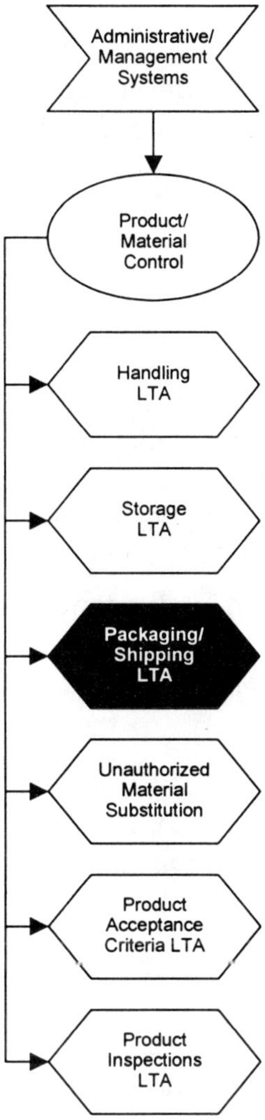

Typical Issues

Was material packaged properly? Was it damaged because of improper packaging? Was equipment exposed to adverse conditions because the packaging had been damaged? Was the material transported properly? Was it damaged during shipping?

Typical Recommendations

- Inspect materials for damage upon initial arrival at the facility
- Ensure that packaging specifications are documented, communicated, and clearly understood by the vendor
- Provide directions for unpacking items so they are not damaged by the customer
- Ensure that proper packaging methods are used for the final product

Examples

An electronic system incurred water damage because it was not packaged in waterproof packaging as specified in the packaging requirements.

An electronic device used for chemical analysis provided incorrect analysis results. As a result, 10,000 gallons of product were later found to be unacceptable. Investigation revealed that the electronic device had been dropped off of a forklift. Because there was no obvious physical damage, the manufacturer shipped the device.

A water-based coating material was peeling off within several days of being applied. This shipment of the coating material had frozen during transport by truck. Freezing changed the adhesiveness of the coating material.

Motorcycle windshields were packaged in cardboard boxes that were held shut with large metal staples. If the staples were not completely pulled out of the box, they would scratch the plastic windshield when it was removed from the package, making the windshield unusable.

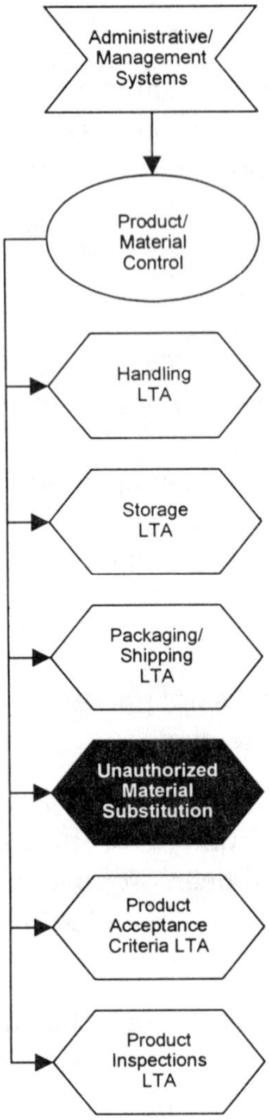

Typical Issues

Were incorrect materials substituted? Were material or parts substituted without authorization? Did the requirements specify no substitution? Did substitution of different materials adversely affect the quality of the final product?

Typical Recommendations

- Implement a management of change program
- Train employees to use the management of change system
- Ensure that field/warehouse personnel understand the management of change system's importance to them
- Assess the impact of material substitutions on the quality of the product produced
- Ensure that materials are properly labeled to prevent inadvertent substitution
- Attempt to design the process so that only the correct item will fit

Examples

A valve failed, causing a spill to the environment. The valve was not the one specified for this service. Because the specified one was not available, a substitute had been installed without the proper review and authorization.

A drawing indicated that 2" bolts would be needed to install a bracket. When the mechanic actually performed the installation, 2.5" bolts were needed. Instead of going back to the warehouse to get the right size bolts, the mechanic used the 2" bolts with the nuts only partially threaded and no lock washer. Later, the bracket fell off when the nuts vibrated off the bolts.

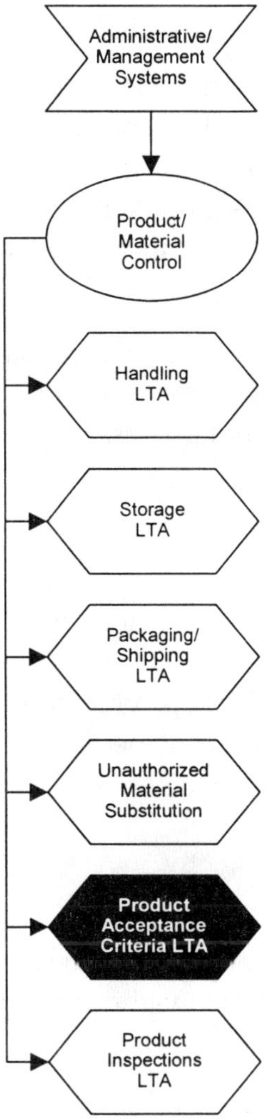

Typical Issues

Could the acceptance criteria for manufactured parts and finished products be understood and implemented? Were the acceptance criteria clear and unambiguous?

Note: *This node only applies to product (things made within your facility) acceptance criteria. Acceptance criteria for purchased materials (items received from outside your facility) are specified as part of the procurement process and are addressed under* Material Acceptance Requirements LTA (Procurement Control).

Typical Recommendations

- Develop acceptance criteria for manufactured parts, materials, and finished products
- Ensure that product acceptance tests can be reasonably implemented

Example

Inspection of toilet paper rolls included checks of the dimensions of the roll, the adequacy of the paper rolling process, and the fragrance added to the roll. Acceptance criteria were specified for the roll dimensions and adequacy of the rolling process. No acceptance criteria existed for the adequacy of the fragrance level. As a result, some batches were shipped without the required fragrance.

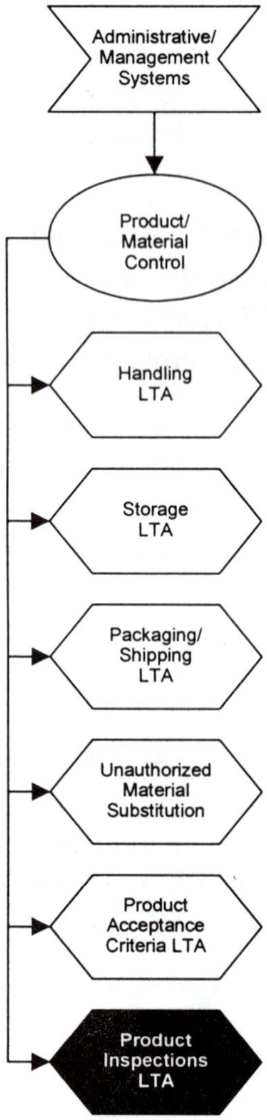

Typical Issues

Was inspection of materials performed in accordance with the acceptance requirements? Did a lack of inspection lead to safety, reliability, or quality problems? Were manufactured parts, materials, and final products inspected prior to shipment? Can the inspection requirements be reasonably implemented? Do intellectual products (i.e., reports, analyses, data) meet requirements?

Note: *This node only applies to materials and work products produced within your facility. Inspections of all materials and work products received from outside the facility are addressed as part of the procurement process under* Material Inspections LTA (Procurement Control).

Typical Recommendations

- Ensure that material/product inspections are performed in accordance with requirements
- Provide clear inspection specifications and methods for product testing
- Provide personnel with the capability to implement the inspection requirements

Examples

Acceptance criteria specified that a moisture test should be performed on a sample of each shipment of powder. The warehouse was not told who was supposed to do the test. As a result, the material was shipped without the test being performed.

Product inspection requirements specified that 10% of all items be inspected before shipment. When one of the quality assurance inspectors was gone (e.g., sick, in training), only 5% could be inspected without holding up shipments. As a result, a number of bad lots of material were shipped.

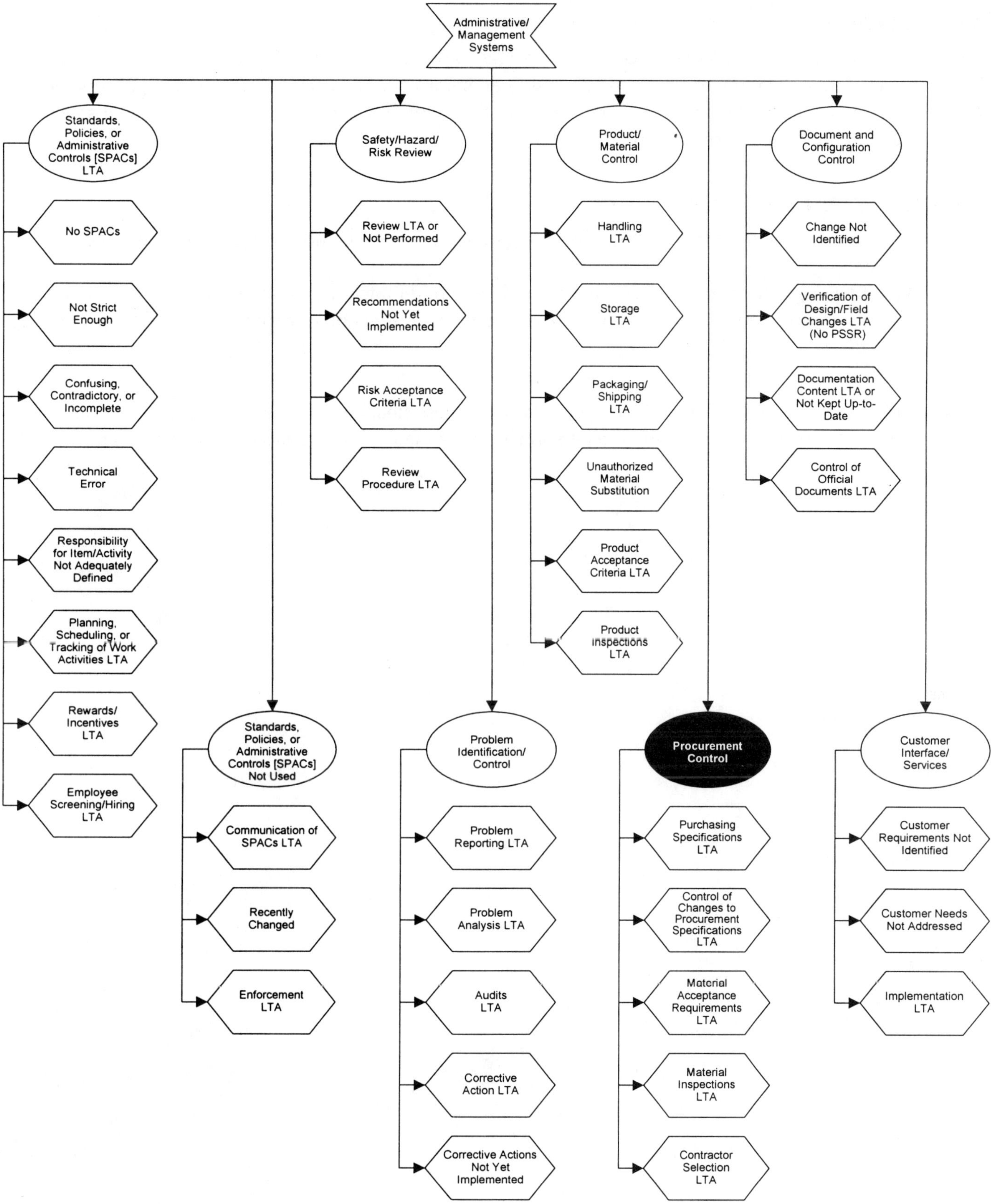
Administrative/ Management Systems
Standards, Policies, or Administrative Controls [SPACs] LTA
No SPACs
Not Strict Enough
Confusing, Contradictory, or Incomplete
Technical Error
Responsibility for Item/Activity Not Adequately Defined
Planning, Scheduling, or Tracking of Work Activities LTA
Rewards/ Incentives LTA
Employee Screening/Hiring LTA
Safety/Hazard/ Risk Review
Review LTA or Not Performed
Recommendations Not Yet Implemented
Risk Acceptance Criteria LTA
Review Procedure LTA
Product/ Material Control
Handling LTA
Storage LTA
Packaging/ Shipping LTA
Unauthorized Material Substitution
Product Acceptance Criteria LTA
Product Inspections LTA
Document and Configuration Control
Change Not Identified
Verification of Design/Field Changes LTA (No PSSR)
Documentation Content LTA or Not Kept Up-to-Date
Control of Official Documents LTA
Standards, Policies, or Administrative Controls [SPACs] Not Used
Communication of SPACs LTA
Recently Changed
Enforcement LTA
Problem Identification/ Control
Problem Reporting LTA
Problem Analysis LTA
Audits LTA
Corrective Action LTA
Corrective Actions Not Yet Implemented
Procurement Control
Purchasing Specifications LTA
Control of Changes to Procurement Specifications LTA
Material Acceptance Requirements LTA
Material Inspections LTA
Contractor Selection LTA
Customer Interface/ Services
Customer Requirements Not Identified
Customer Needs Not Addressed
Implementation LTA

Typical Issues

Was the error the result of inadequate control of changes to procurement specifications or purchase orders? Did a fabricated item fail to meet requirements or was an incorrect item received? Did product acceptance requirements fail to match design requirements or were they otherwise unacceptable? Were proper specifications and evaluations used to select contractors?

Typical Recommendations

- Procurement specifications should not be changed without review and approval by knowledgeable personnel
- Ensure that acceptance requirements are documented and match the design requirements
- Ensure that the contractor selection process considers the impact on overall cost, reliability, and quality

Example

A large tank was fabricated using an incorrect grade of stainless steel because the buyer made an unauthorized change to the purchase order, and the personnel who signed off on the order did not detect the change.

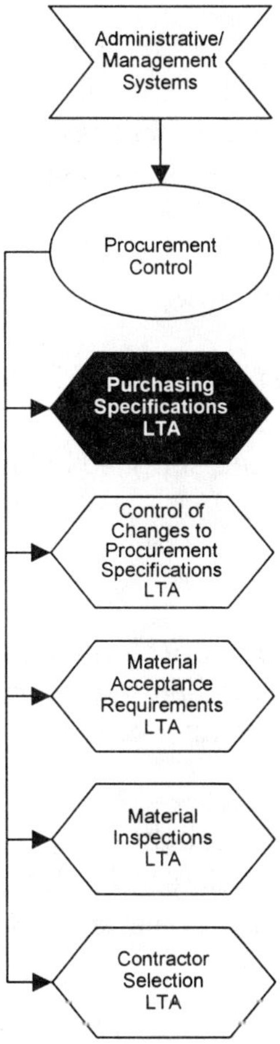

Typical Issue

Did the purchase specifications include (1) a schedule for delivery of the materials, (2) material packaging and shipping requirements, (3) safety requirements, (4) liability clauses, and (5) payment schedules?

Note: *This node applies to HOW items are obtained, not WHAT is obtained.* See Material Acceptance Requirements LTA *for problems related to specification of what was purchased.*

Typical Recommendation

- Develop purchase specifications with input from the technical contacts, procurement specialists, attorneys, and others in your company to ensure that all contractual requirements are addressed

Example

A contract to purchase logs from a supplier did not include late delivery penalties. As a result, the supplier was routinely a week or two behind schedule.

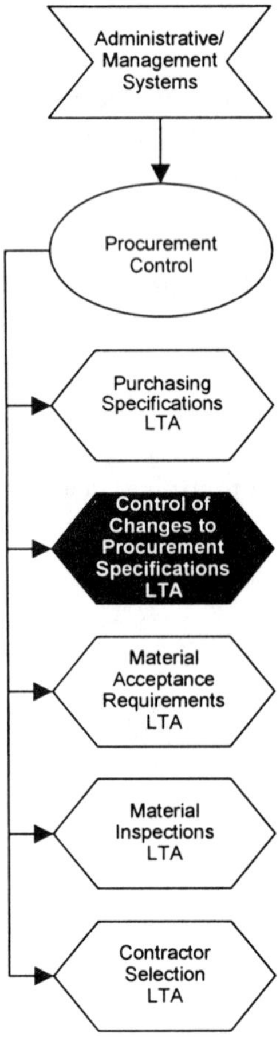

Typical Issues

Were changes made to purchase orders or procurement specifications without the proper reviews and approvals? Did the changes result in purchase of the wrong materials? Did changes in contract language cause safety, reliability, quality, or legal problems?

Typical Recommendations

- Include procurement control procedures in the management of change program
- Provide receipt inspection that compares the materials supplied against the original plant request

Examples

A pump made from Hastelloy C was ordered for use in a hypochlorite liquid plant. Purchasing went out for bids on Hastelloy pumps (and did not specify Hastelloy C). A Hastelloy B pump was received, and failed after only 4 days of service because of chemical attack.

During the purchasing process, the procurement specialist bought unformatted diskettes because they were cheaper. As a result, each disk had to be formatted by the user. This resulted in about 40 hours of wasted production time while the disks were formatted.

A contract to hire subcontractors originally required the contractors to supply hazardous material handling training to their personnel at the contractor's expense. This requirement was subsequently dropped. As a result, the company had to pay for the training and pay the contractor for the time their personnel spent in the training.

A batch of product was ruined because of improper mixing of the components. Purchasing had switched suppliers to reduce costs. The feed material was now purchased at twice the concentration as before. The management of change system did not identify it as a change because the same material was purchased from both suppliers.

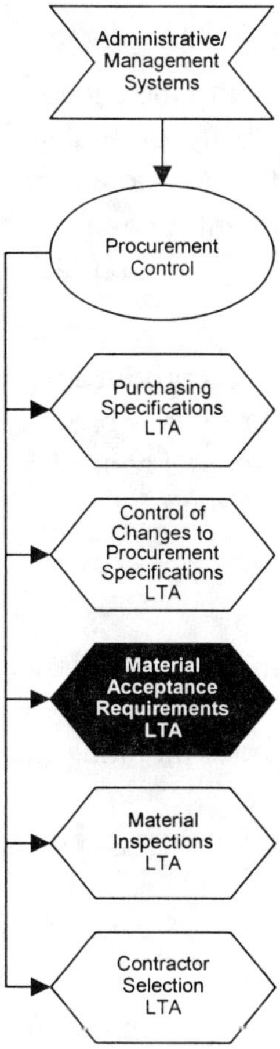

Typical Issues

Were acceptance criteria for raw materials, spare parts, and process equipment adequate? Was it easy to determine if the material received was acceptable?

Note: *This node applies to problems related to WHAT was purchased. Problems associated with the process of obtaining, paying for, and delivering the material is covered under* Purchasing Specifications LTA.

Typical Recommendation

- Develop acceptance criteria for raw materials, spare parts, and process equipment
- Have the warehouse personnel assist in the development of the acceptance criteria to ensure that they are clearly understood by those who will use them

Examples

Acceptance criteria specified that the bolts should have a Rockwell-C hardness of 30. Warehouse personnel did not know what this meant or how to determine if the bolts met this specification.

Acceptance criteria specified that the powder should not contain excessive moisture. Warehouse personnel did not know exactly what this meant. As a result, they accepted material that was unusable.

Acceptance criteria had not been developed for rubber gaskets used in a process. The gaskets deteriorate rapidly if they are not individually sealed in plastic. Without any acceptance criteria, the warehouse accepted a shipment of gaskets that were not individually wrapped and sealed.

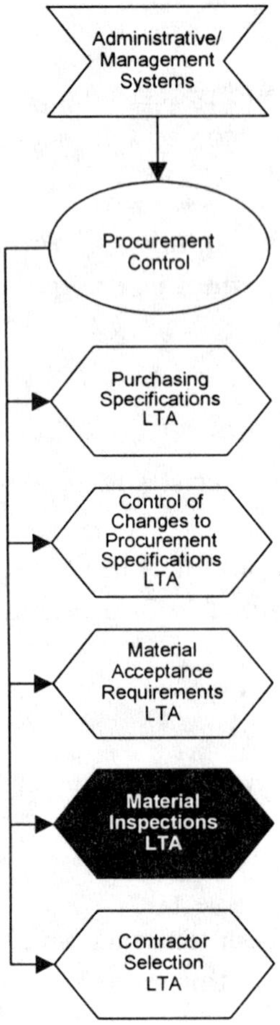

Typical Issues

Was the inspection of materials performed in accordance with the acceptance requirements? Did a lack of inspection lead to safety, reliability, or quality problems? Were manufactured parts or materials inspected prior to acceptance or use in your process? Can the inspection requirements be reasonably implemented? Do intellectual products (i.e., reports, analyses, data) meet requirements?

Note: *This node only applies to materials and work products produced outside your facility. Inspections of all materials and work products produced within your facility are addressed as part of the product material control process under* Product Inspections LTA (Product/Material Control).

Typical Recommendations

- Ensure that material receipt inspections are performed in accordance with requirements
- Provide clear inspection specifications and methods for material acceptance.
- Provide personnel with the capability to implement the inspection requirements

Examples

Acceptance criteria specified that a moisture test should be performed on a sample of each shipment of powder received from a supplier. The warehouse was not told who was supposed to do the test. As a result, the material was accepted and used without the test being performed.

The acceptance criteria for a raw material specified that a lengthy test be performed before the material would be transferred from the tanker to the supply tanks. Another, less rigorous, test was often substituted to save time.

An outside contractor was employed to perform a safety analysis of a system in accordance with OSHA requirements. No one reviewed the contractor's report prior to closing out the contract. The contractor failed to analyze all of the portions of the system that were in the scope of the contract.

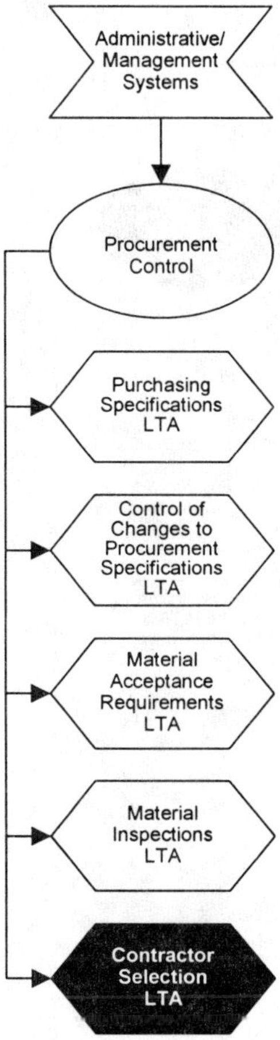

Typical Issue

Does the contractor selection process address the following: (1) safety requirements, (2) training, (3) liability, and (4) scheduling?

Typical Recommendation

- Develop purchase specifications for contract services with input from the technical contacts, procurement specialists, attorneys, and others in your company to ensure that all contractual requirements are addressed

Examples

A contract to hire subcontractors did not specify who was responsible for paying for hazardous material handling training for the contract personnel. As a result, the company had to pay for the training and pay the contractor for time their personnel spent in the training.

The contract for supplying maintenance personnel did not specify that equipment supplied and used by the contractor be subject to approval by the company. As a result, the contractor used ladders and other equipment that did not meet OSHA requirements.

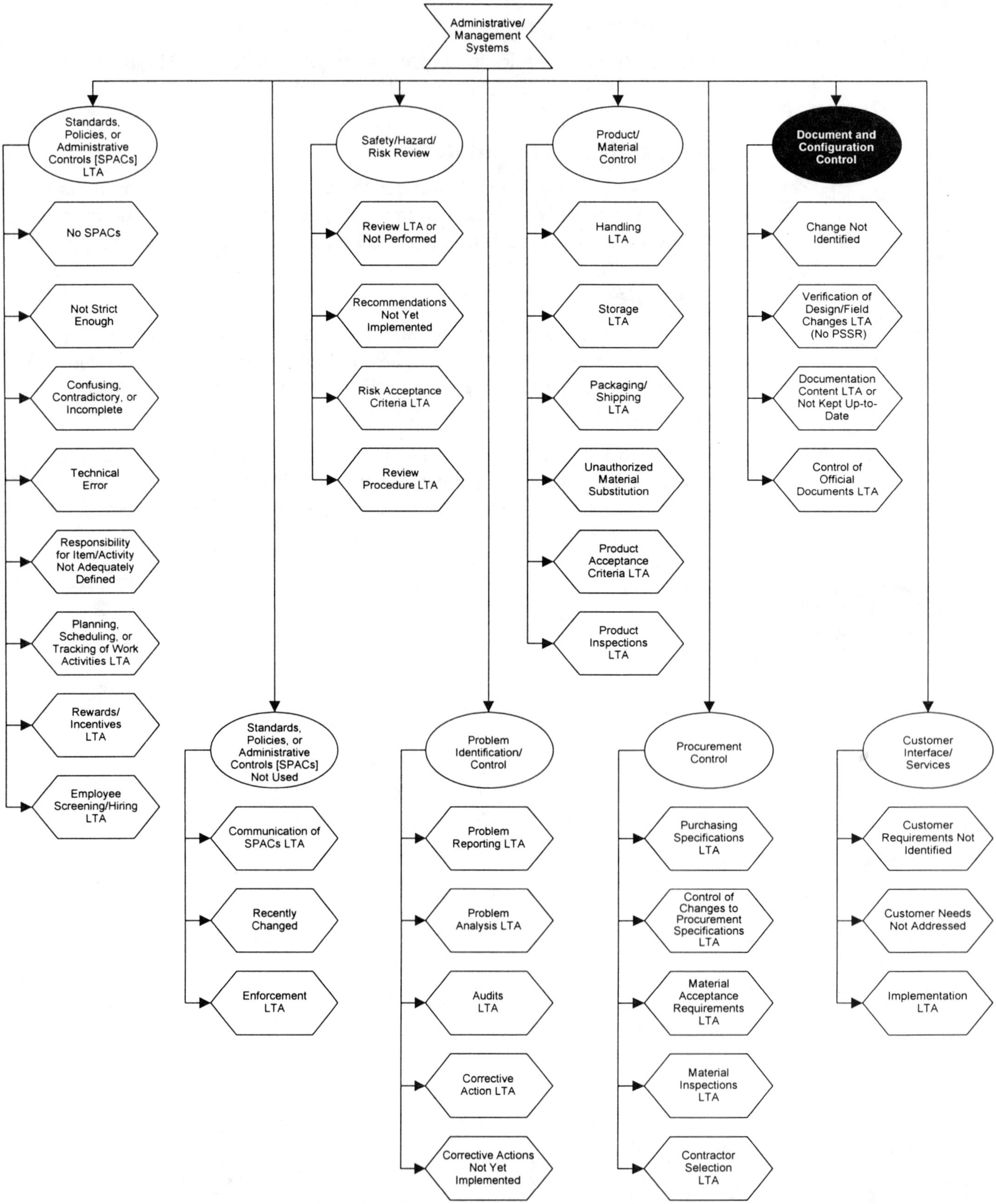
Administrative/ Management Systems
Standards, Policies, or Administrative Controls [SPACs] LTA
No SPACs
Not Strict Enough
Confusing, Contradictory, or Incomplete
Technical Error
Responsibility for Item/Activity Not Adequately Defined
Planning, Scheduling, or Tracking of Work Activities LTA
Rewards/ Incentives LTA
Employee Screening/Hiring LTA
Standards, Policies, or Administrative Controls [SPACs] Not Used
Communication of SPACs LTA
Recently Changed
Enforcement LTA
Safety/Hazard/ Risk Review
Review LTA or Not Performed
Recommendations Not Yet Implemented
Risk Acceptance Criteria LTA
Review Procedure LTA
Problem Identification/ Control
Problem Reporting LTA
Problem Analysis LTA
Audits LTA
Corrective Action LTA
Corrective Actions Not Yet Implemented
Product/ Material Control
Handling LTA
Storage LTA
Packaging/ Shipping LTA
Unauthorized Material Substitution
Product Acceptance Criteria LTA
Product Inspections LTA
Procurement Control
Purchasing Specifications LTA
Control of Changes to Procurement Specifications LTA
Material Acceptance Requirements LTA
Material Inspections LTA
Contractor Selection LTA
Document and Configuration Control
Change Not Identified
Verification of Design/Field Changes LTA (No PSSR)
Documentation Content LTA or Not Kept Up-to-Date
Control of Official Documents LTA
Customer Interface/ Services
Customer Requirements Not Identified
Customer Needs Not Addressed
Implementation LTA

Typical Issues

Were drawings or documentation not complete or up-to-date? Was control of design/field changes inadequate? Was the error caused by improper control of as-built documents? Was an "unofficial" copy of a procedure/drawing used?

Typical Recommendations

- Train all employees to understand the difference between a change and a replacement-in-kind.
- Field changes should be reviewed and approved
- Periodically audit to verify that all official copies are updated

Examples

A batch of product was ruined because of improper mixing of the components. Purchasing had switched suppliers to reduce costs. The feed material was now purchased at twice the concentration as before. The management of change system did not identify it as a change because the same material was purchased from both suppliers.

An operator was using an out-of-date process drawing in the field because it contained all of his markups. The markups were required to correct errors on the drawing and to add additional information the drawing did not contain.

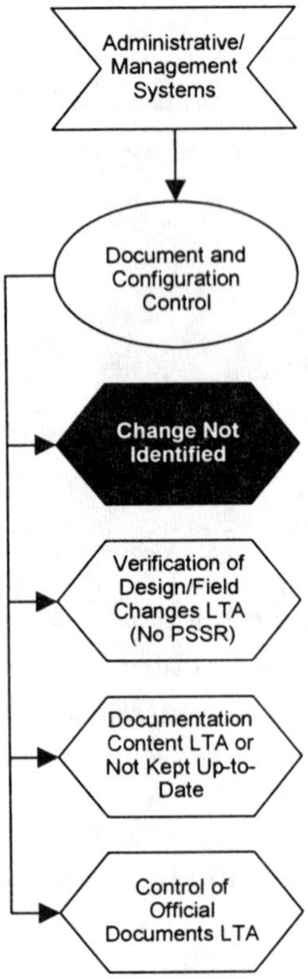

Typical Issues

Was the change to the system identified? Was the definition of change LTA? Did personnel understand the definition of "change" versus "replacement-in-kind"?

Typical Recommendations

- Ensure that authorization signatures are obtained from key personnel before design/field changes can be implemented
- Train all employees to understand a change versus a replacement-in-kind
- Train employees on how to initiate a request for change
- Provide specific examples of what is and is not a change requiring review

Examples

An SO_2 release occurred because a stiffer gasket was installed. The gasket installed could last longer in this chemical service, but would not seal properly using previous torque settings. The management of change system defined "replacement-in-kind" as use of "similar or better" materials. Because the maintenance department considered the new gasket material superior, a change review was not performed.

A new supplier was selected to supply product barrels to the facility. Barrels from the new supplier were cheaper but only came in one color (black). This caused shipment problems because different colored barrels had been used previously to easily identify the barrel contents. Purchasing did not realize the importance of the color coding.

A field modification to an instrument air line had to be made to route the line around a water line that was not on the drawings used by the designer. This reroute created a low point in the air line where contaminants collected. The field modification was not identified as a change that required a review.

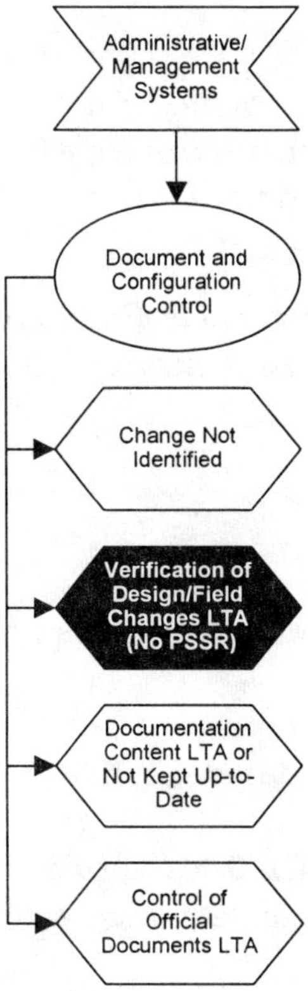

Typical Issues

Were new equipment or installations verified to conform to specifications prior to startup? Were new or modified components functionally tested prior to startup?

Typical Recommendations

- Conduct a pre-startup safety review for new or modified facilities, and ensure that all requirements of the review have been met before highly hazardous chemicals are introduced into the process
- Conduct an assessment of field changes and new installations to ensure proper operation of the equipment following startup

Examples

A control valve failed to the wrong position upon loss of instrument air. A pre-startup safety review was not performed because the valve was installed as part of a replacement-in-kind.

A new air compressor was installed. A pre-startup review of the installation was performed to ensure that it was installed correctly. However, no operational tests of the compressor were performed. As a result, the compressor failed soon after startup because of an insufficient cooling water supply.

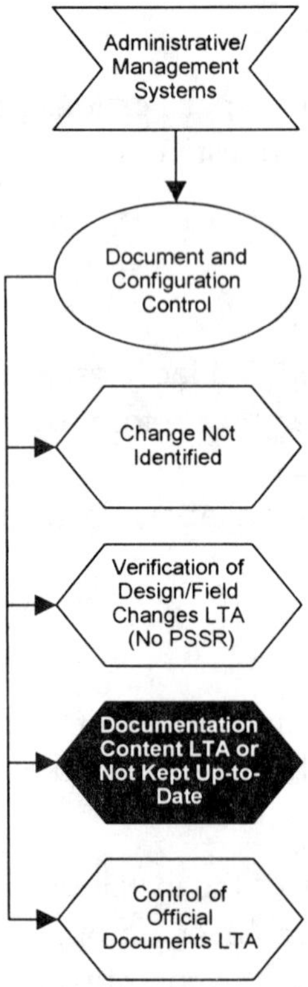

Typical Issues

Were drawings and documents updated when changes were made? Did documents/drawings reflect the current status? Do documents contain all of the required information? Do documents used in the field have markups to make them useful?

Note: *This node applies primarily to drawings and operator aids. Procedure content problems are addressed under* Procedures. *Problems associated with the content of SPACs are addressed under* SPACs LTA.

Typical Recommendations

- Require authorization signatures for all design/field changes
- Include the task of updating drawings and procedures in the document change tracking system
- Solicit input from document users on required changes.
- Involve the document users in periodic reviews and updates of the documents
- Consider conducting mandatory walkthroughs in the field following construction to confirm that official documents accurately reflect actual design

Examples

A settling tank was moved 4 feet from its original location to allow for proper forklift access to other equipment. This field change was not indicated in the final design documentation. As a result, a skid-mounted demineralizer installation had to be field modified because the settling tank took up part of the floor to be used for installation of the demineralizer skid.

Two system modifications were being implemented concurrently; however, the design engineers did not know this. The drawings did not indicate that changes were pending from these two modifications. As a result, changes implemented by the first modification were undone by implementation of the second modification.

An acid spill occurred during opening of a line break. Lockouts had been made based on current drawings. The drawings were not up-to-date and did not show an acid stream that had been tied into the line 3 months earlier. The system that existed for controlling documents was not adequate. The organization was 6 months behind on updating marked-up drawings and distributing new copies to all official document holders.

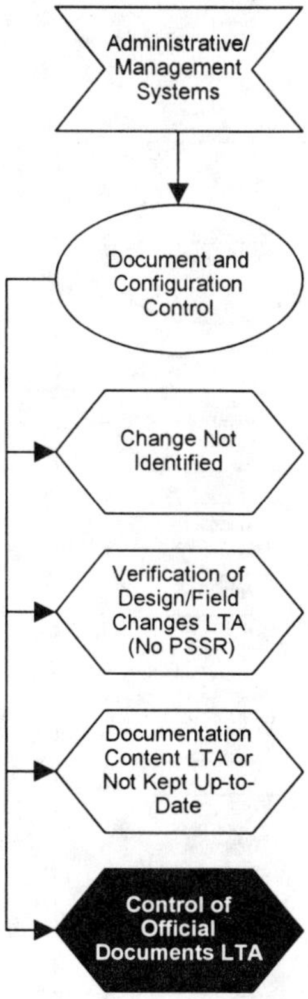

Typical Issues

Did a system exist for controlling documents? Did the system provide methods for keeping documents up-to-date? Were all necessary documents available? Were all official copies of each document updated? Were all unofficial copies or outdated copies found or disposed of? Are procedures, standards, policies, and other official documents used in the field current?

Note: *This node applies to the distribution of documents. Problems related to the content of the document are addressed under* Documentation Content LTA or Not Kept Up-to-Date.

Typical Recommendations

- "Search and destroy" unofficial copies of documents
- Periodically conduct an audit to ensure that all official copies are updated

Example

Maintenance personnel often made printouts of procedures that they kept at their workstations. That way they did not need to get a new copy of the procedure each time. However, they did not check for updates each time before use.

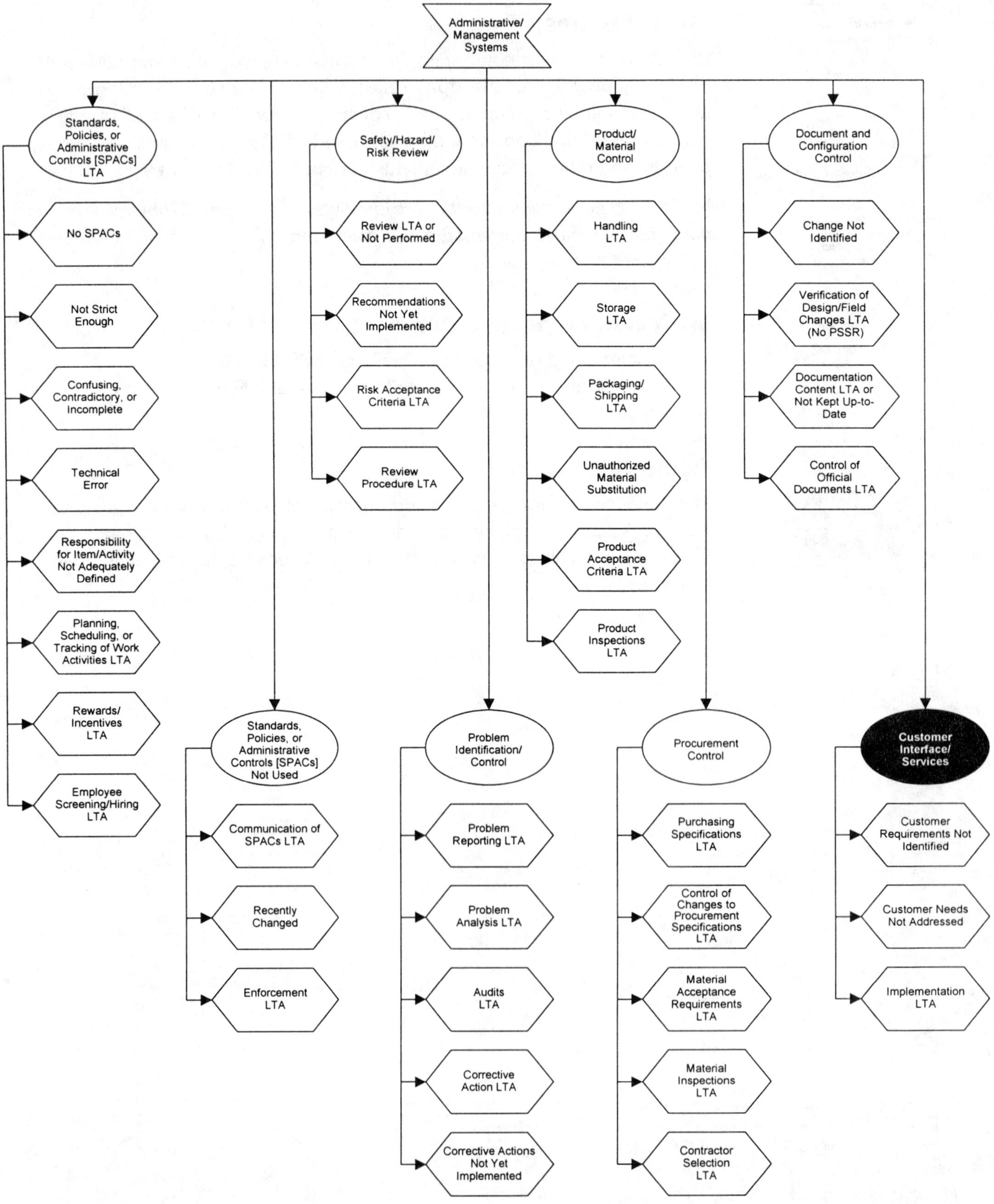
Administrative/ Management Systems
Standards, Policies, or Administrative Controls [SPACs] LTA
No SPACs
Not Strict Enough
Confusing, Contradictory, or Incomplete
Technical Error
Responsibility for Item/Activity Not Adequately Defined
Planning, Scheduling, or Tracking of Work Activities LTA
Rewards/ Incentives LTA
Employee Screening/Hiring LTA
Standards, Policies, or Administrative Controls [SPACs] Not Used
Communication of SPACs LTA
Recently Changed
Enforcement LTA
Safety/Hazard/ Risk Review
Review LTA or Not Performed
Recommendations Not Yet Implemented
Risk Acceptance Criteria LTA
Review Procedure LTA
Problem Identification/ Control
Problem Reporting LTA
Problem Analysis LTA
Audits LTA
Corrective Action LTA
Corrective Actions Not Yet Implemented
Product/ Material Control
Handling LTA
Storage LTA
Packaging/ Shipping LTA
Unauthorized Material Substitution
Product Acceptance Criteria LTA
Product Inspections LTA
Procurement Control
Purchasing Specifications LTA
Control of Changes to Procurement Specifications LTA
Material Acceptance Requirements LTA
Material Inspections LTA
Contractor Selection LTA
Document and Configuration Control
Change Not Identified
Verification of Design/Field Changes LTA (No PSSR)
Documentation Content LTA or Not Kept Up-to-Date
Control of Official Documents LTA
Customer Interface/ Services
Customer Requirements Not Identified
Customer Needs Not Addressed
Implementation LTA

Typical Issues

Are there problems associated with the customer interface? Are customer needs identified? Are customer needs addressed?

Typical Recommendations

- Develop a system to solicit feedback from customers
- Develop a system to allow customers to easily contact your company
- Ensure that customer requests are promptly addressed

Examples

A customer required special product packaging to allow it to quickly load the product into its system. Some shipments were made using the special container while others were not. This requirement was not always passed on to the plant personnel.

Customers ordered materials from a catalog. Not all items in the catalog were being manufactured. Customer service still accepted orders for these items because they were not told that they were no longer being made.

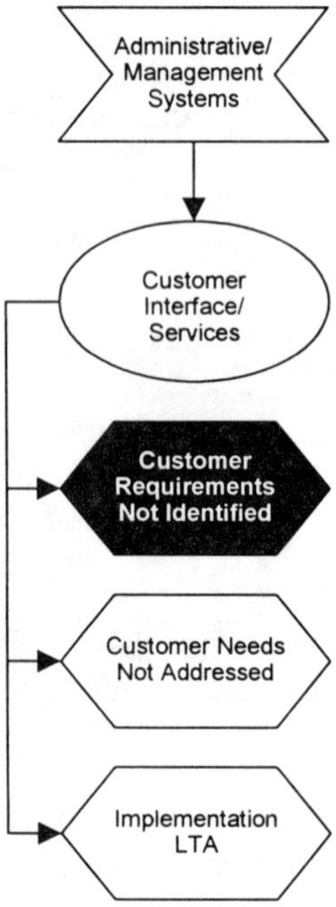

Typical Issues

Are customer requirements for products identified? Is there a mechanism in place to solicit customer requirements?

Typical Recommendations

- Provide incentives to personnel to solicit and clarify customer requirements
- Provide a means to handle special orders and requests

Examples

A customer initially ordered 20,000 pounds of material A. Later, the customer changed the order to 30,000 pounds of material B. This change was never input into the order system. As a result, the wrong material was delivered to the customer.

A printer used a standard blue color to produce pamphlets for a client. A customized blue color should have been used to match other advertising materials used by the client. The printer just assumed that a standard color (which is less expensive) was what the client wanted.

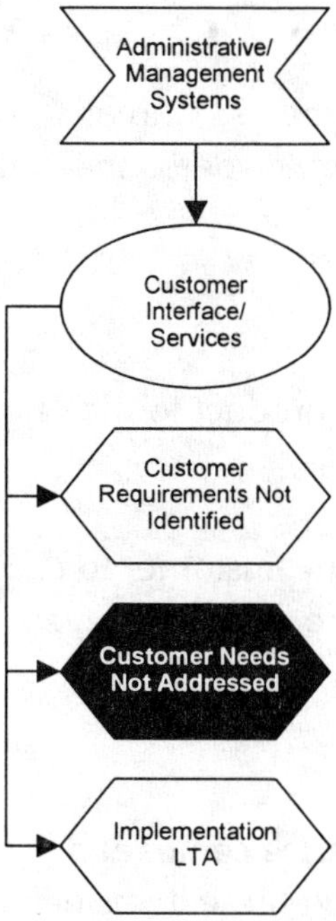

Typical Issues

Are customer needs passed on through the organization to those who need to know about them? Are there methods to produce products that meet unusual requirements?

Typical Recommendations

- Ensure that customer requirements are passed on to all those in the organization who need to know about them
- Provide a means to flag special orders to make them easy to identify

Example

A customer required special product packaging to allow it to quickly load the product into its system. Some shipments were made using the special container while others were not. This requirement was not always passed on to the plant personnel.

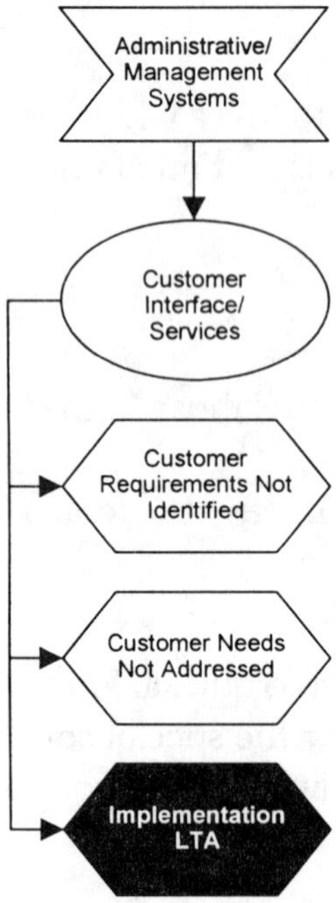

Typical Issues

Were there problems in addressing the customer requirements? Can all customer requirements be addressed? Were customer service personnel helpful? Were customer service personnel courteous?

Typical Recommendations

- Provide a method to close the loop. Compare the product to the original customer requirements
- Develop a system to track special orders
- When problems are encountered, follow up with the customer to clarify what went wrong. Ensure that follow-up actions are taken to prevent recurrence

Example

Special customer orders were taken by the customer service center and passed on to the plant. The plant placed a low priority on these items because they would interrupt normal production. As a result, most special orders were delivered several weeks after the promised delivery date.

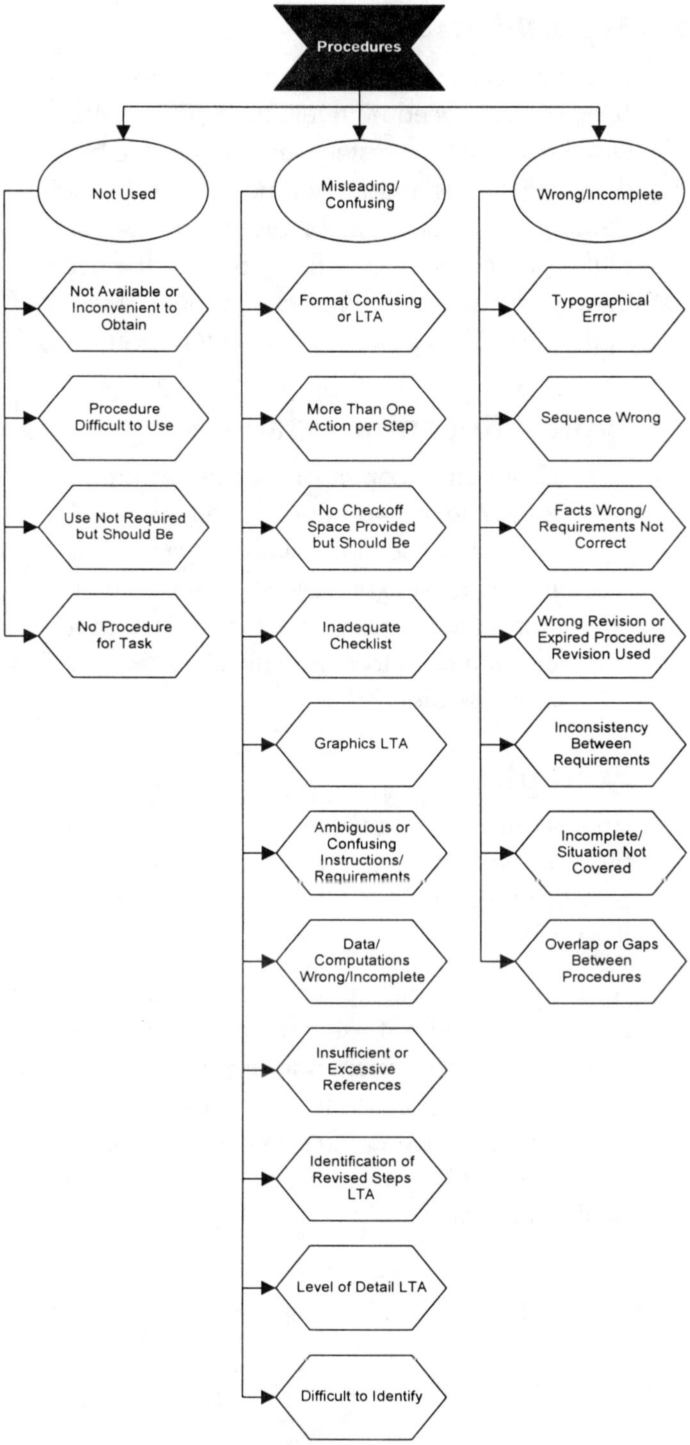

Typical Issues

Was a procedure used to perform the job? Was the procedure incorrect or incomplete? Was a procedure developed for the job? Was a procedure required to perform the job?

Note: *Procedures provide detailed, step-by-step directions on how to accomplish a task. Guidance documents that provide general guidance and principles should be addressed under* SPACs LTA *or* SPACs Not Used (Administrative/Management Systems).

Typical Recommendations

- Ensure that copies of procedures are available for worker use at all times
- Ensure that procedures are in a standard, easy-to-read format
- Perform a walkthrough of new and revised procedures
- Use look-up tables instead of requiring calculations to be performed

Examples

An operator failed to complete a critical step in an operation because the procedure he obtained from the procedure files was not the most recent revision.

A new operator failed to complete a critical step because the procedure was not detailed enough; it was written as a guideline/reminder for experienced operators.

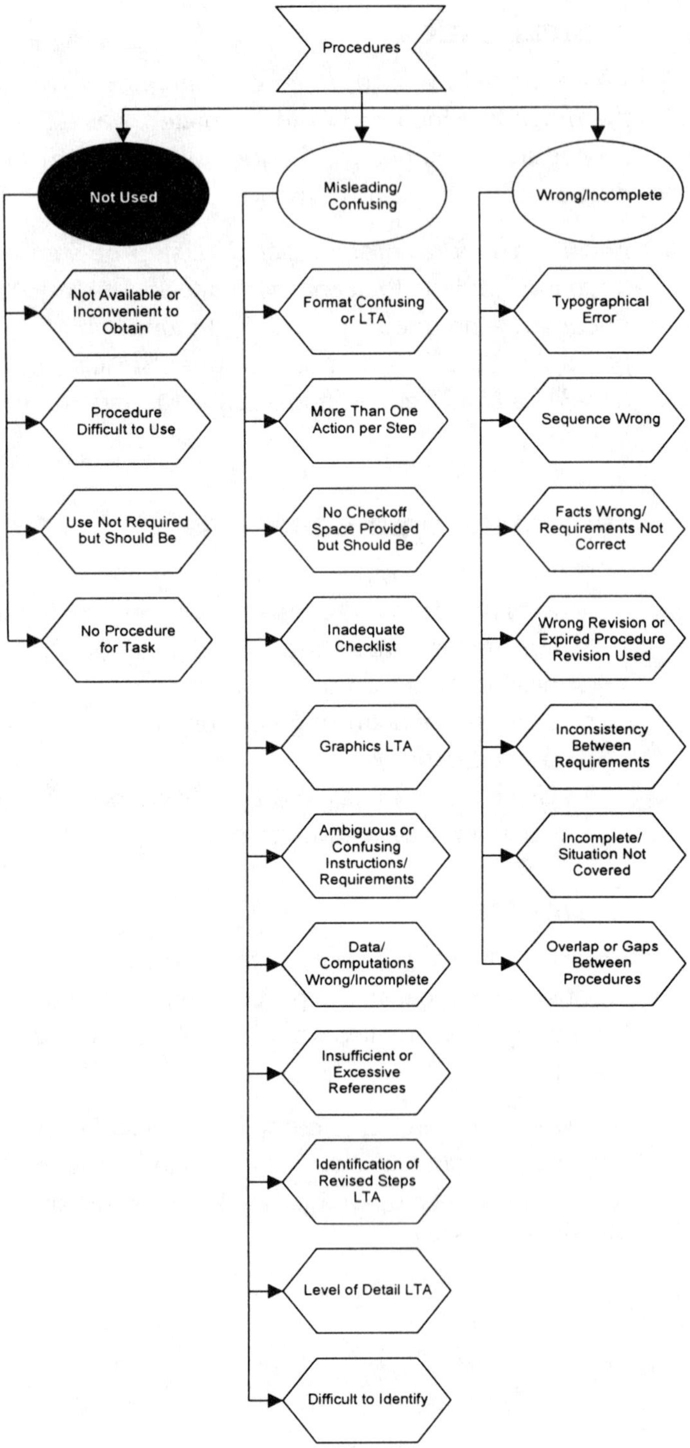

Typical Issues

Was a procedure used to perform the job? Was a copy of the procedure available to the worker? Did the procedure system require that the procedure be used as a task reference or was it just for training? Were personnel required to take copies of the procedures to the field? Should the use of the procedure be required even though it was not in the past? Was a procedure written for this task?

Typical Recommendations

- Ensure that copies of procedures are available for worker use at all times
- Develop procedures with sufficient detail for the least experienced, qualified worker
- Supplement training and reference materials with easy-to-carry checklists that parallel a procedure

Examples

An operator made a valving error. He performed the task without using the controlled procedure because he would have had to make a copy of the master.

A mechanic incorrectly performed a repair job on a key pump without using the procedure. Mechanics were not required to use the procedure in the field because it was for training purposes only. However, using the procedure in the field would probably have prevented the error made by the mechanic.

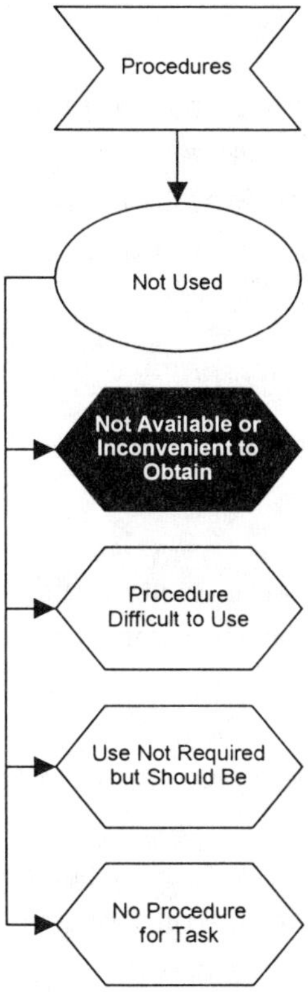

Typical Issues

Did a procedure exist for the job or task being performed? Was the procedure readily available? Was there a copy of the procedure in the designated file, shelf, or rack? Was there a "master copy" of the procedure available for reproduction?

Typical Recommendations

- Place copies of operations and maintenance procedures in the appropriate work areas so that the procedures are ALWAYS available for personnel use
- Maintain master copies of all procedures and control access to these masters

Examples

An operator made a valving error. He did not use the controlled procedure because it would have required him to make a copy of the master. Instead he used the procedure copy he had at his workstation. This procedure was out-of-date.

An electrician was troubleshooting a large breaker. After determining what the problem was, she should have obtained a copy of the procedure for replacement of the charging springs. But that would have required her to return to the maintenance shop. So she replaced the spring based on memory. As a result, a plant startup was delayed when the breaker failed to close.

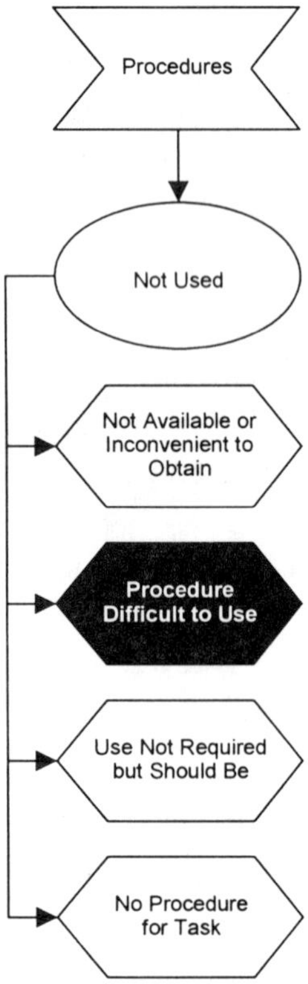

Typical Issues

Considering the training and experience of the user, was the procedure too difficult to understand or follow? Was sufficient information available to identify the appropriate procedure? Was the procedure designed for the "less practiced" user? Was procedure use inconvenient because of working conditions (e.g., tight quarters, weather, protective clothing)?

Note: *Dual coding under* Misleading/Confusing (Procedures) *may also be appropriate.*

Typical Recommendations

- Develop procedures such that the content provides the least experienced employee with adequate direction to successfully complete required tasks
- Choose a procedure format that is easy to read and follow
- Choose a procedure format that is appropriate to the level of complexity of the task
- If certain job tasks require an employee to be in an awkward position or to wear uncomfortable personal protective equipment, make procedure use as convenient as possible by posting applicable procedures at eye level in an easy-to-read format in these specific locations
- If tasks require reference in the field to a procedure, ensure that employees are provided with a concise yet complete (with no references to other procedures) procedure (or checklist) that is easy to carry and use in the field (like a one- or two-page printout of the pertinent procedure)

Examples

An inexperienced mechanic made a mistake installing a piece of equipment. The mechanic did not take a copy of the procedure with him because it was long, it used terminology that he did not understand, and he felt he understood the task well enough.

An inexperienced mechanic made a mistake installing a piece of equipment. The procedure stated only to remove the old item and replace it with a similar unit. This was not detailed enough for the inexperienced mechanic.

The operator did not use the procedure because of its numerous cross-references to other procedures. To carry all of them would have required a large notebook.

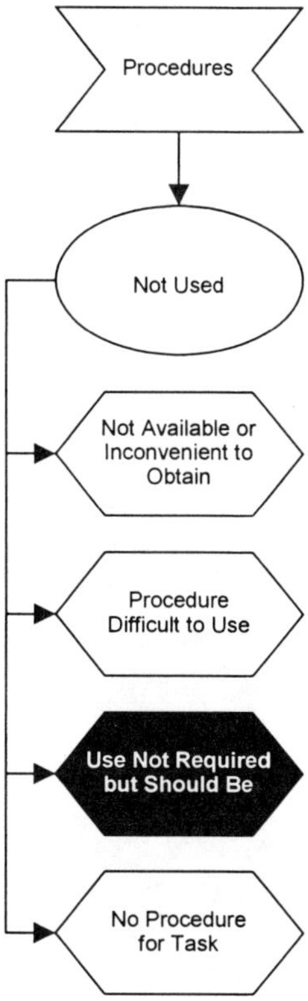

Typical Issues

Was the procedure classified for training and reference? Based upon the significance or difficulty of the job, should the procedure have been classified as a "use every time" procedure?

Typical Recommendations

- Procedures classified as reference procedures should contain very few steps. If the number of steps is too overwhelming for short-term memory, it should be classified as a "use every time" procedure
- Training and reference manuals may need to be supplemented by:
  - easy-to-carry checklists that parallel the procedure
  - more detailed step-by-step procedures for "use every time" if the training and reference manuals are too cumbersome

Examples

An operator made a valving error, resulting in a tank overflow. He did not take a copy of the procedure with him because it was for reference, and he thought he knew how to perform the valving operation.

A mechanic incorrectly performed a repair job on a key pump without using the procedure. Mechanics were not required to use the procedure in the field because it was for training purposes only. However, using the procedure in the field would probably have prevented the error made by the mechanic.

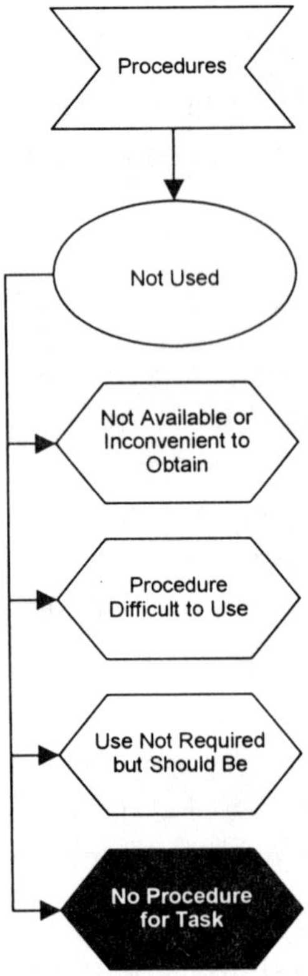

Typical Issue

Was there a procedure for this task?

Typical Recommendations

- Develop a procedure for the task
- Ensure that all modes of operation, all maintenance activities, and all special activities have written procedures

Example

A mechanic under-torqued a flange. He performed the job without a procedure because one did not exist for the task.

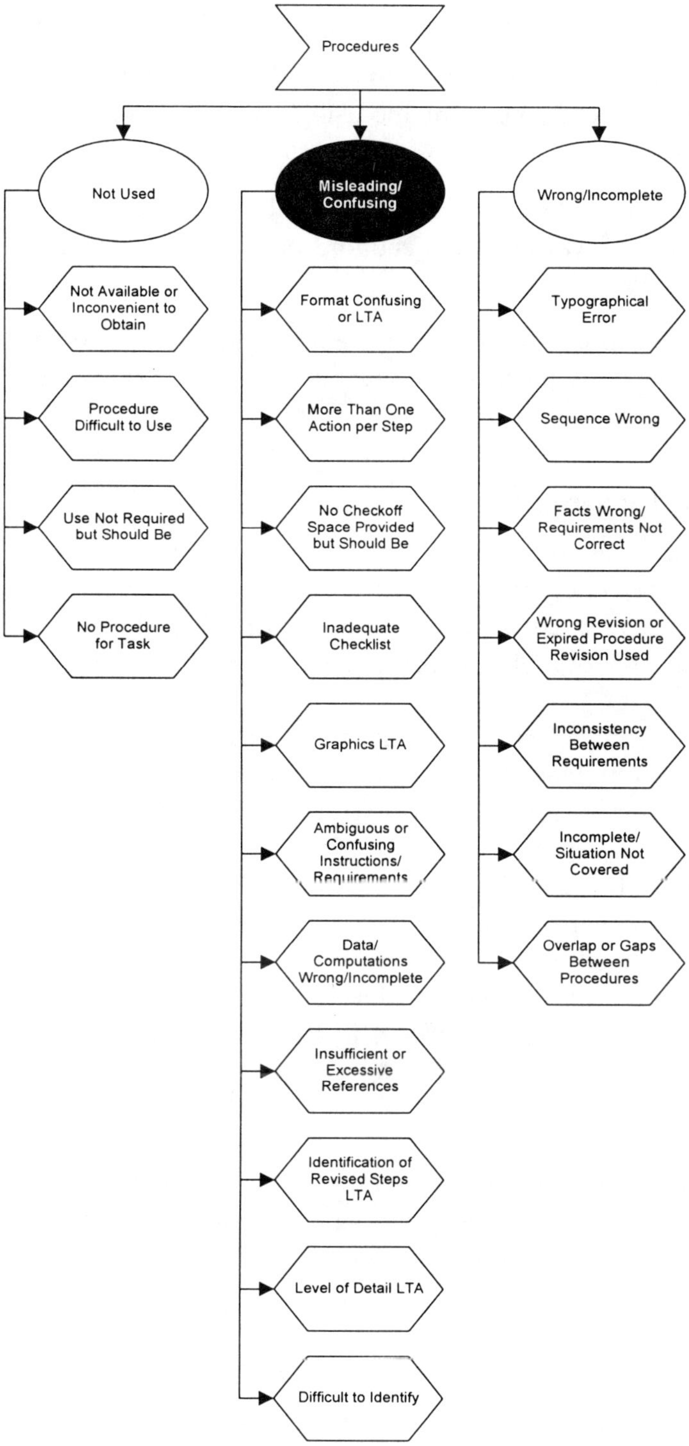

Typical Issues

Was an event caused by an error made while following or trying to follow a procedure? Was the procedure misleading or confusing?

Typical Recommendations

- Ensure that procedures are in a standard, easy-to-read format
- Ensure that procedures use the appropriate level of detail for the complexity and frequency of a task
- Use look-up tables instead of requiring calculations to be performed
- Use specific component identifiers

Examples

An operator incorrectly completed a step of a procedure requiring him to open six valves. He skipped one of the valves. The procedure did not have a checkoff space for each valve.

An operator overfilled a tank. The procedure required him to calculate the running time of the fill pump. A look-up table with the initial tank level and the corresponding fill pump run time should have been provided.

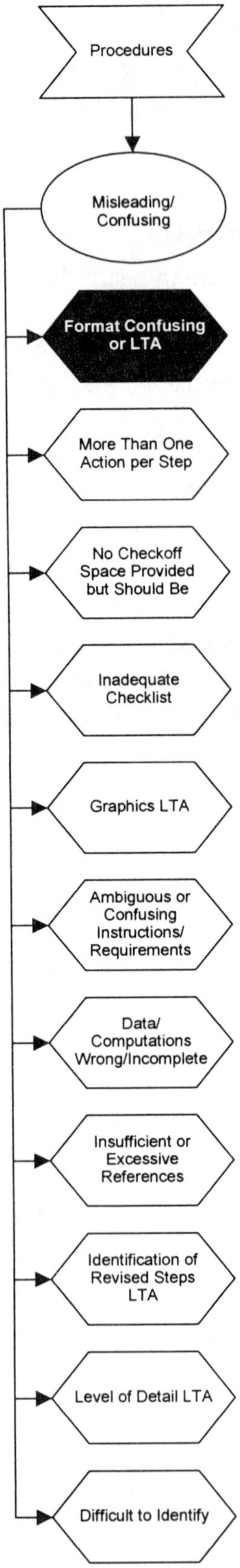

Typical Issues

Did the layout of the procedure make it difficult to follow? Did the format differ from that which the user was accustomed to using? Were the steps of the procedure logically grouped?

Do warnings or cautions contain information that should be contained in procedure steps? Are important warnings and cautions embedded in procedure steps?

Is the procedure format appropriate for the task? Is a flow chart used when a checklist is more appropriate? Is a checklist used when a T-bar format is more appropriate?

Typical Recommendations

- Ensure that procedures are in an easy-to-read format. Use color codes (or change paper color) when appropriate.
- Avoid using the narrative or paragraph format; personnel tend to get lost in a sea of print. The T-bar, flowchart, or checklist formats are highly effective
- Choose one or two effective formats and use these same formats consistently throughout the facility. The format for a troubleshooting guide may be inappropriate for a step-by-step startup procedure
- List procedure steps in a logical, sequential order. Also, be sure that any special precautions are listed at the beginning of the procedure
- Review procedures to ensure that warnings and cautions are presented in a consistent format in all procedures
- Involve procedure users in the procedure development process. Have an inexperienced user review the procedure to ensure that sufficient detail is provided
- Use checklists for verification processes and initial alignments of systems
- Use flowcharts when decisions affect which part of the procedure is implemented (e.g., a troubleshooting guide, or an emergency procedure that requires diagnosis of the problem)

Examples

An operator made a mistake while performing a startup procedure. The procedure was confusing because it required the operator to complete part of section A, then B, back to A, then to C, back to A, then to D and E. The operator failed to go back to A after completing C.

Each step in the procedure was numbered. Subsequent levels of substeps were numbered by adding a decimal point and another set of numbers. The procedure used too many levels on substeps (i.e., a step was numbered 2.3.6.5.1.1.1.1.5). As a result, the operator skipped a step in the procedure.

A troubleshooting guide was developed using a checklist format. The mechanics did not understand how to move through the procedure; they just completed the items they thought were appropriate.

A procedure was developed by an engineer in a paragraph format. About half of the information in the procedure was design information that the operators did not need.

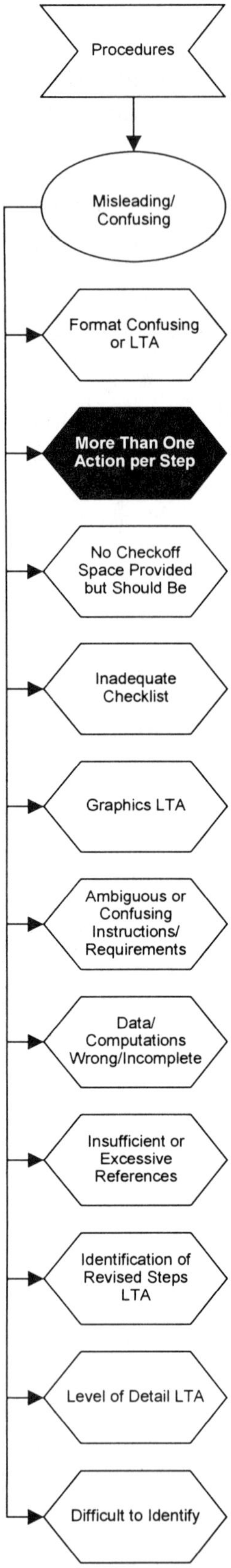

Typical Issues

Did any steps in the procedure have more than one action or direction to perform? Did some steps in the procedure state one action, which, in practice, actually required several steps to perform?

Typical Recommendations

- Avoid broad procedure steps such as "Charge the reactor." Instead, use this as a subheading and include all the steps associated with charging the reactor below the heading
- Do not assume that an employee will remember all the steps associated with an action item. Clearly communicate all the required steps associated with an action item so that the least experienced employee can successfully perform the required job tasks

Example

An operator failed to close a valve, resulting in a tank overflow. The instruction to close the valve was one of six actions required in one step of the procedure. He completed the other five actions but overlooked closing the valve, which was the fourth action in the step.

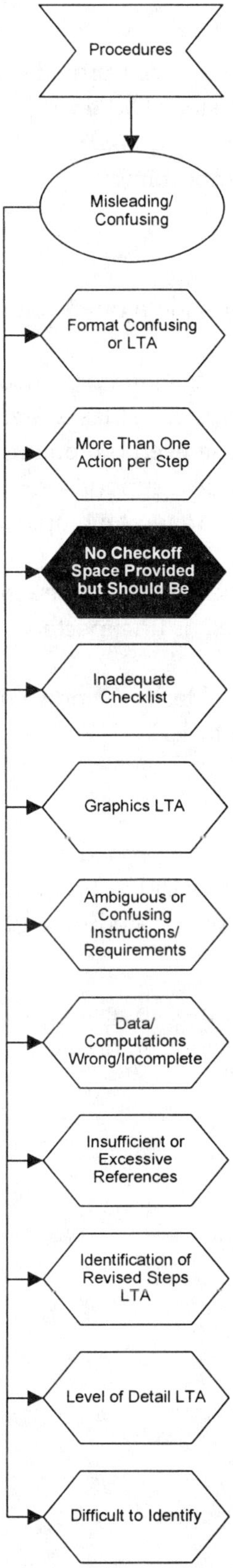

Typical Issues

Was an error made because each separate action in a step did not have a checkoff space provided? Is the procedure complex and critical enough to require checkoffs?

Typical Recommendations

- For actions that require multiple steps, ensure that all the steps are specifically defined. When appropriate, include a checkoff space for each of these individual steps so that the employee can be certain that he/she has performed this step
- It is a good practice to design procedures with enough "white space" (by indentation, line spacing, etc.) to allow users to keep their place when using the procedure

Example

An operator failed to open a valve. The procedure required him to open seven valves. He missed one, opening the other six. A separate checkoff space for each valve manipulation was not provided in the procedure.

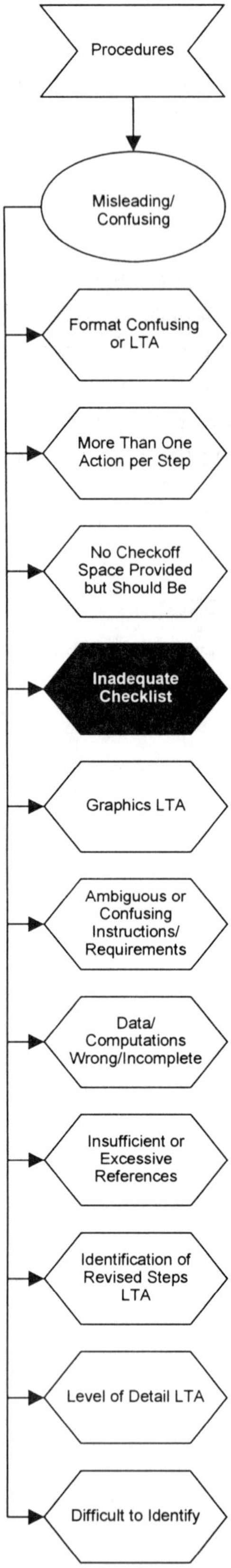

Typical Issues

If a checklist was necessary, was it confusing? Was enough room provided for the response or did it require unique responses for each step? Did each instruction (regardless of format) clearly indicate what was required? Was a detailed checklist required for a task that was not very important?

Typical Recommendations

- Develop a checklist for all safety-critical tasks to provide a quick reference for inexperienced and experienced users
- Require that checklists be turned in if necessary for quality assurance
- Avoid using checklists instead of supervision to ensure that tasks are performed correctly because checklists can easily be filled out before or after the task; if supervision is required, then provide a supervisor
- Include the unique system response to be expected when an employee completes each step of a checklist
- Provide enough white space on the checklist so that the employee may record the system response so that expected as well as unexpected responses can be documented
- Ensure that checklists are only developed for critical tasks. Overuse of checklists will reduce their effectiveness on critical tasks

Examples

An operator failed to complete one step of a procedure. The procedure required a check at the completion of each step. Because it did not require unique responses for the steps, the operator completed the procedure and then checked off all the steps at one time.

A checklist was designed so that the desirable answer to most questions (23 out of 26) was yes. As a result, the three remaining questions were often answered incorrectly.

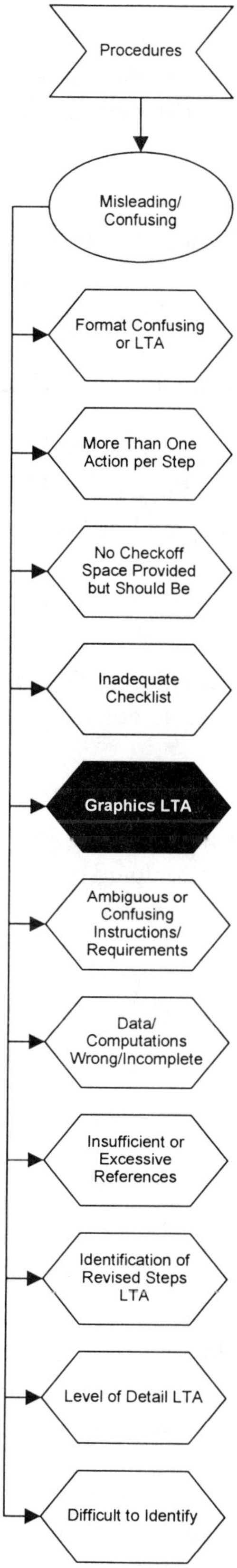

Typical Issues

Was an error made because graphics or drawings were of poor quality? Were the graphics or drawings unclear, confusing, or misleading? Were graphics, including data sheets, legible? Would a graphic (diagram, picture, chart, etc.) have made a significant reduction in the likelihood of this error were it provided?

Typical Recommendations

- For hard-copy graphics that have been reproduced, ensure that the copy is easy to read (e.g., not too dark, too light, or splotchy)
- Include color coding on graphics when possible for easy use
- Ensure that the graphics accurately depict actual process operations and/or equipment configuration
- Do not overwhelm the user with too many graphics on one screen or one sheet of paper. Information should not appear crowded
- The text should support the graphics
- Flowcharts can be very effective graphics for tasks that require decision making and branching

Examples

A mechanic replaced the wrong seal on a large piece of equipment. The seal that he was to remove was shaded on the drawing, but he could not determine which seal was shaded because the copy was of poor quality.

An electrician incorrectly terminated a wire. The wire terminations were shown on the installation diagram. The procedure copy he was using was not legible because it was made from a copy of a copy of a copy of the original.

An operator opened two valves in the wrong sequence during a complex procedure to backwash an enclosed rotary filter containing highly reactive peroxides. A diagram of the filter (showing equipment labels) and proper labeling of the filter valving would have greatly clarified the procedure.

An operator had to determine if the reactor's temperature and pressure were acceptable. The acceptable temperature was dependent on the pressure. The operator had a long set of look-up tables that listed the acceptable temperature for each pressure. A pressure-temperature graph indicating acceptable and unacceptable regions would have reduced errors.

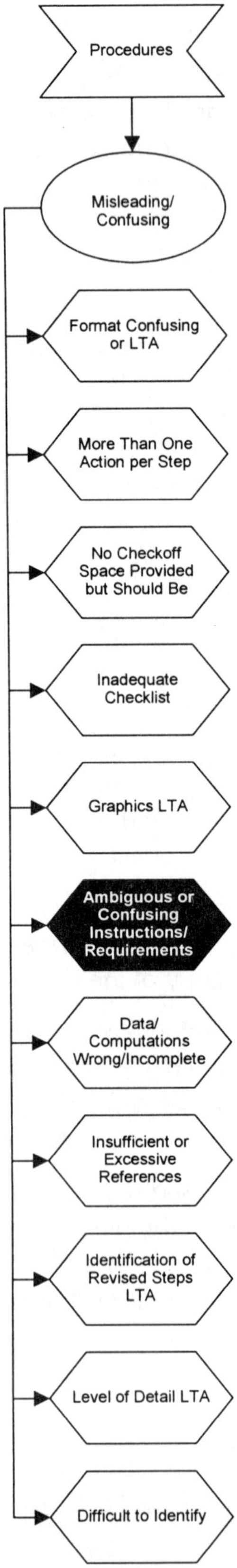

Typical Issues

Were the instructions in the procedure unclear? Could they be interpreted in more than one way? Was the language or grammar unclear/complex?

Typical Recommendations

- Have procedures validated by a team of subject matter experts (workers) and by walkthroughs in the field
- To find difficult steps, have the newest employee walk through the procedure without coaching
- Allow technical editors to review procedures to ensure that ambiguous terms have been avoided
- Perform a hazard review of critical procedures to determine other accident scenarios related to errors in procedures and to determine if sufficient safeguards are provided against employees not following the written procedures

Examples

An instruction called for cutting XYZ rods into 10-foot-long pieces. The intent was to have pieces 10 feet long. The person cutting the pieces cut 10 pieces, each a foot long.

A step in the root cause analysis procedure stated "Use the RCM to assist in determining the management system deficiencies that contributed to the event." The supervisor assumed it meant to use reliability-centered maintenance not the Root Cause Map.

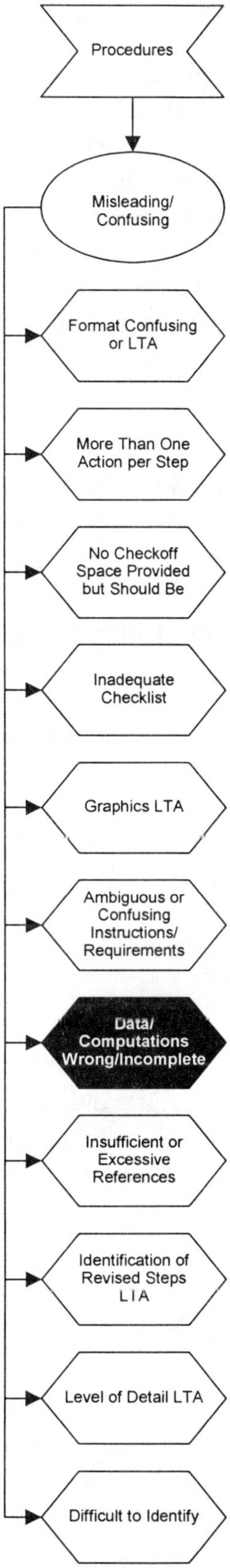

Typical Issues

Was the error made because of a mistake in recording or transferring data? Were calculations performed incorrectly? Was the formula or equation confusing? Did it have multiple steps?

Note: *Consider dual coding with* Ambiguous or Confusing Instructions/ Requirements *or* More Than One Action per Step

Typical Recommendations

- Avoid procedures that require employees to make manual calculations. Instead, provide employees with pre-calculated tables or worksheets with easy-to-fill-in blanks and with thorough training in their use. Or, automate calculations within the system
- Perform a hazard review of critical procedures to determine other accident scenarios related to errors in procedures and to determine if sufficient safeguards are provided against employees not following the written procedures

Example

A procedure required the operator to calculate the weight of product in the tank based on the empty weight of the tank and the current weight of the tank. Both of these values were displayed on the computer. The operator made an error in subtracting the numbers. The computer display could be modified to display this number and eliminate the need for the calculation

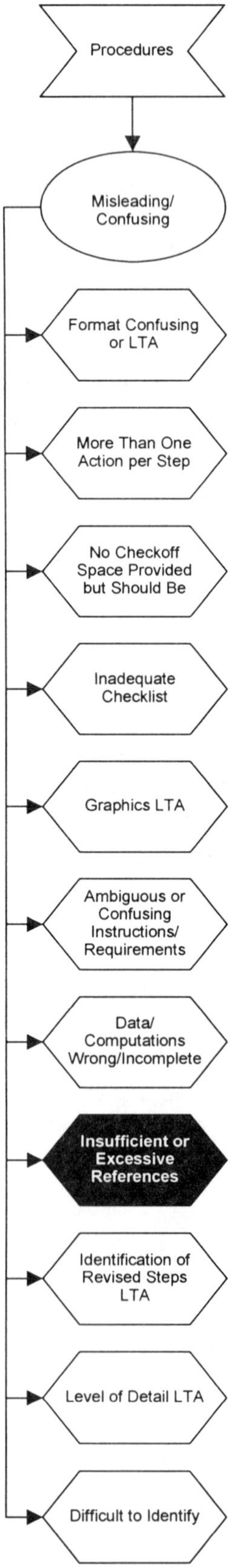

Typical Issues

Did the procedure refer to an excessive number of additional procedures? Did the procedure contain numerous steps of the type "Calculate limits per procedure XYZ"? Was the procedure difficult to follow because of excessive branching to other procedures? Did the procedure contain numerous steps of the type "If X, then go to procedure ABC. If Y, then go to procedure EFG"? Did the procedure contain numerous references to other parts of the procedure? Did it contain steps of the type "If the material is acceptable, go to Step 13.3. If the material is unacceptable, go to Step 12.4. If the test cannot be run, redo Step 4 and contact your supervisor."

Typical Recommendations

- List all information that an employee must have in order to perform a specific task in the procedure designated for this task. If the same information is required to perform different tasks, repeat the information in each procedure
- Do not branch (reference) to more than one other procedure (module) from a procedure
- Procedures intended for step-by-step use in the plant/field need to contain all required tasks; an employee is unlikely to return to the file/manual to get any referenced procedure
- Perform a hazard review of critical procedures to determine other accident scenarios related to errors in procedures and to determine if sufficient safeguards are provided against employees not following the written procedures
- Use a flowchart to determine the correct procedure steps to be implemented. Avoid too many jumps within a procedure

Example

An operator exceeded an operating limit. The primary procedure did not contain the limits but referred to four other procedures to find the limits. When checking his results against the limits, he looked at the wrong limit in one of the referenced procedures.

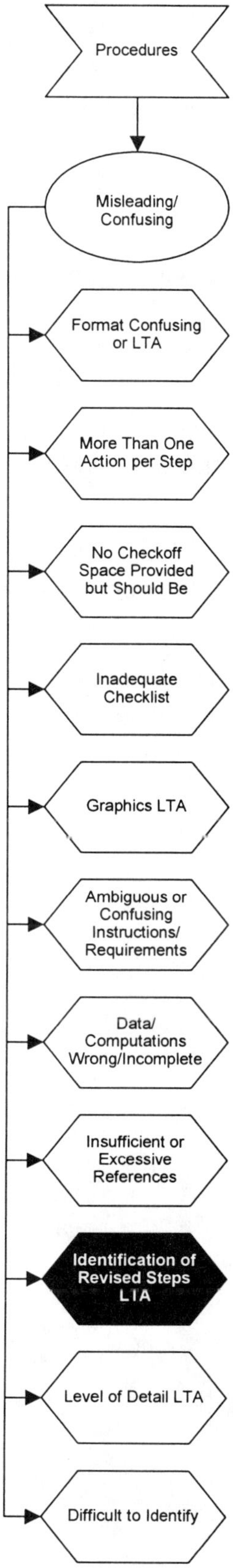

Typical Issues

Was the procedure user required to carry out actions different from those he was accustomed to doing? Did the procedure identify that the step for the action had been revised? Did the procedure user perform the action as the previous revision specified rather than the current revision?

Typical Recommendations

- Ensure that procedure changes are managed, and that all official copies are updated and all unofficial copies destroyed
- Clearly identify (such as with a sidebar) which steps/information have changed, and ensure that all employees are trained in or informed of changes
- Avoid the use of multiple area references within a procedure

Example

An operator incorrectly completed a step of a procedure. The operator was experienced and performed the action as he always had. The new procedure (which had been correctly updated) was not marked to indicate that the step had recently been revised, and the operator did not realize that a change had been made.

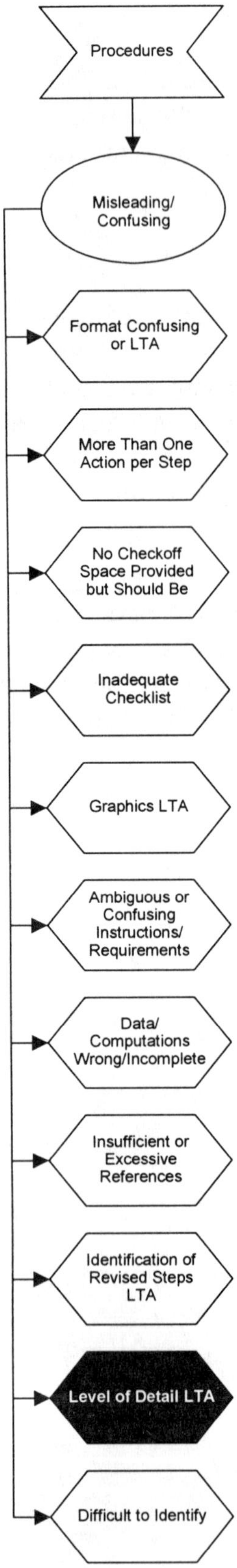

Typical Issues

Do the procedures provide too little detail to ensure proper performance of the task by the most inexperienced operator? Do the procedures have too much detail?

Typical Recommendation

- Consider using an outline format with high level steps for experienced users and detailed steps for inexperienced users

Examples

The instructions for a computer software program just stated "change the loading preferences to user-defined values." No further directions were provided on how this could be done.

An engineer developed a procedure in paragraph format. About half of the information in the procedure was design information that the operators did not need.

An operations procedure for the shutdown of the cooling water system included specific steps on how to close manually operated valves. This information was not needed in the procedure because it was a common operator skill that did not require any task-specific knowledge.

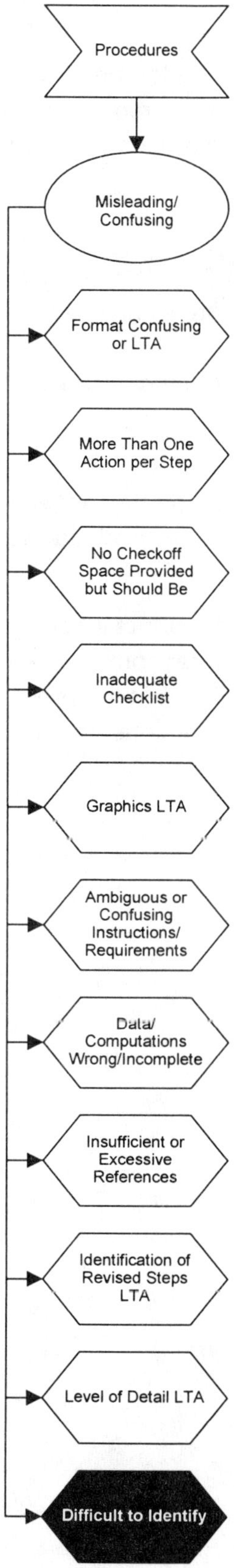

Typical Issues

Is it difficult to identify the correct procedure to use? Do many procedures have similar names? Are the procedures for different units clearly distinguishable from one another?

Typical Recommendations

- Include a header at the top of each procedure page that includes the procedure number, page number, procedure revision, and unit number
- Use different colored paper for each unit's procedures (i.e., blue for Unit 1, pink for Unit 2)
- Provide clear, descriptive names for each procedure

Examples

An operator used the wrong procedure to start up compressor 3A. There were three procedures labeled "Startup of Compressor 3" (for compressors 3A, 3B, and 3C). The procedure he used was for compressor 3C.

A mechanic incorrectly calibrated a pressure transmitter. A page from a similar procedure was inadvertently substituted into his calibration procedure. Individual procedure pages did not contain procedure titles or procedure numbers, so the substituted page was difficult to distinguish from the others.

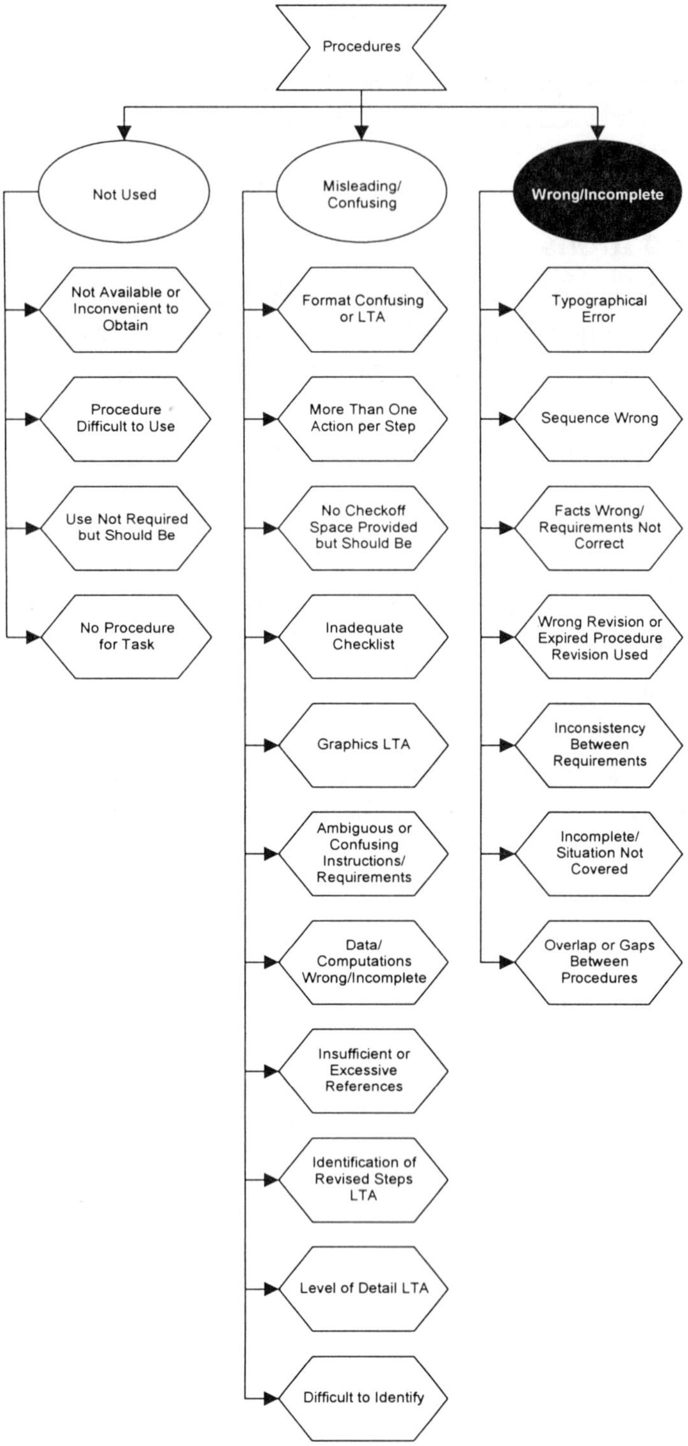

Typical Issues

Was the procedure incorrect? Did the procedure fail to address a situation that occurred during performance of the task? Is the procedure consistent with the installed equipment?

Typical Recommendations

- Ensure that procedures are technically reviewed
- Perform a walkthrough of procedures

Examples

A mechanic made a mistake calibrating a piece of equipment because the procedure specified the wrong limits.

An operator ruined a batch of product when he incorrectly operated the computer control system. New software had been installed, but the procedure had not been updated to be consistent with the new software.

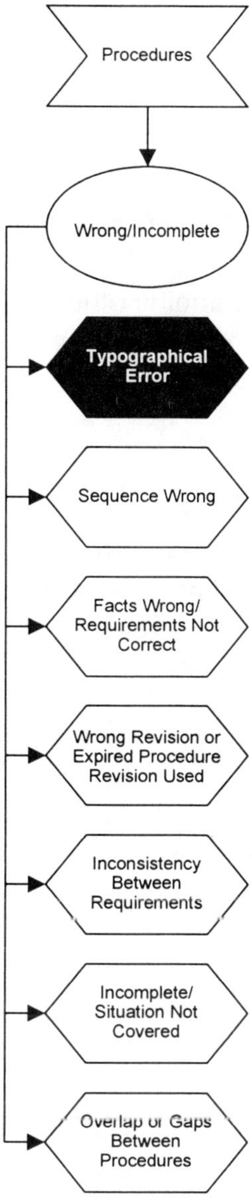

Typical Issue

Was a typographical error in the procedure responsible for the event?

Typical Recommendations

- Use a word processor to electronically spell-check the procedure immediately after it has been typed
- Allow a technical editor to review procedures for typographical errors
- Allow employees to review procedures for accuracy. Solicit feedback from employees

Examples

An operator made a mistake because the procedure contained the wrong limit. The maximum temperature was supposed to be 38° C, but the procedure said 48° C. The mistake was made during typing and not caught by the validators.

An operator overfilled a tank because of a procedure error. The procedure should have stated "Hold the valve open for 3-4 seconds." The typist inadvertently removed the hyphen (when the spell-checker in the word processing software flagged this potential misspelling) and the procedure then read, "Hold the valve open for 34 seconds."

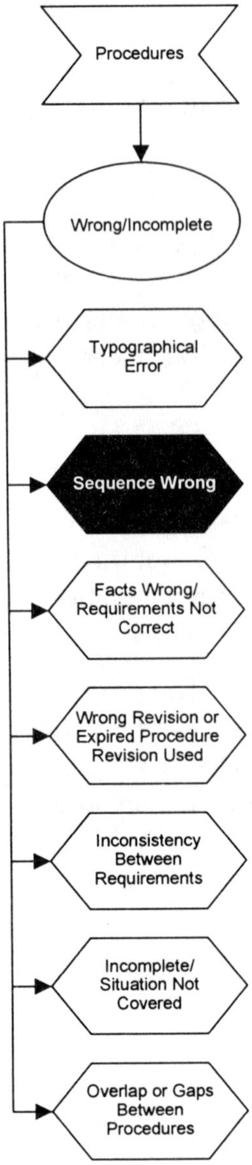

Typical Issue

Were the instructions/steps in the procedure out of sequence?

Typical Recommendations

- Have procedures validated by a team of subject matter experts (workers) and by walkthroughs in the field
- Perform a hazard review of critical procedures to determine other accident scenarios related to errors in procedures and to determine if sufficient safeguards are provided against employees not following the written procedures

Example

An operator made a mistake because the steps were out of sequence in a procedure. Step 5 directed the operator to transfer material from Tank A to Tank B. Step 7 directed the operator to sample the contents of Tank A before transferring.

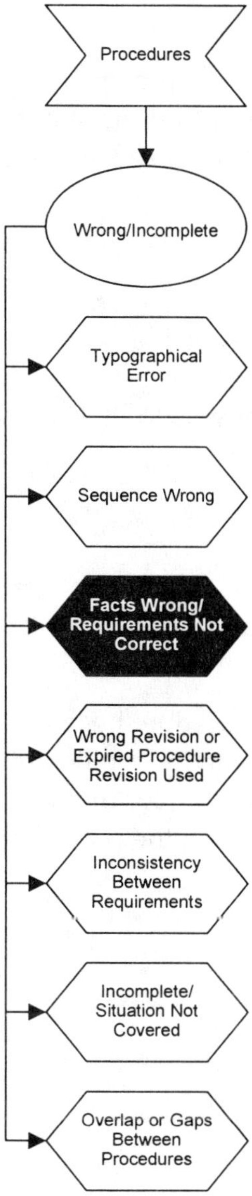

Typical Issues

Was specific information in the procedure incorrect? Did the procedure contain current requirements? Did the procedure reflect the current status of equipment?

Typical Recommendations

- Have procedures validated by a team of subject matter experts (workers) and by walkthroughs in the field
- Perform a hazard review of critical procedures to determine other accident scenarios related to errors in procedures and to determine if sufficient safeguards are provided against employees not following the written procedures

Example

A safety limit was violated because the procedure did not contain the current limits. The limits had been changed, but the master procedure had not been revised.

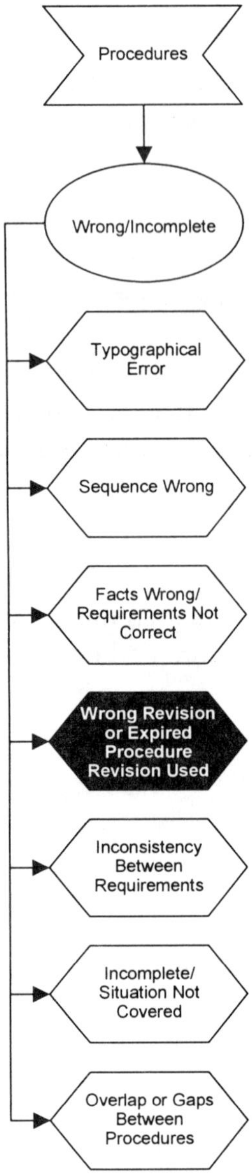

Typical Issues

Was specific information in the procedure incorrect? Did the procedure contain current requirements? Did the procedure reflect the current status of equipment? Was an older version of the procedure used because it was too difficult to obtain a current copy?

Typical Recommendations

- Ensure that only current copies of procedures are available
- Seek out and destroy old versions of the procedures
- Consider incorporating information added by operators to their "personal" copies of procedures

Examples

A safety limit was violated because the procedure did not contain the current limits. The limits had been changed, but the master procedure had not been revised.

An operator liked to use his marked-up version of the procedure because it contained the system operating limits, which were contained in a different procedure. The operator always checked his personal version for updates, but he missed adding a recent change. As a result, he shut down the process when he performed the procedure incorrectly.

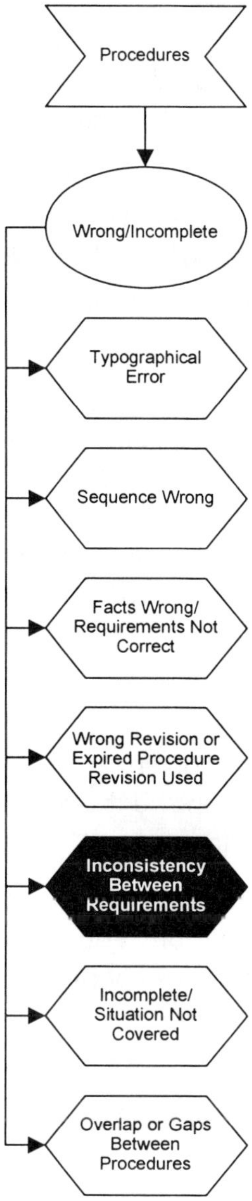

Typical Issues

Did different procedures related to the same task contain different requirements? Were there conflicting or inconsistent requirements stated in different steps of the same procedure? Were requirements stated in different units?

Typical Recommendations

- Have procedures validated by a team of subject matter experts (workers) and by walkthroughs in the field
- Perform a hazard review of critical procedures to determine other accident scenarios related to errors in procedures and to determine if sufficient safeguards are provided against employees not following the written procedures

Examples

An operator exceeded the environmental discharge limits. A caution in the procedure stated the flow rate limit in pounds per hour of material. The procedure step stated the limit in gallons per minute. The operator set the flow rate based on the gallons per minute limit, which was less restrictive in this case.

The procedure said to send the completed form to the PSM Coordinator, but the form itself had a note on the bottom that said to send it to the operations manager.

A caution stated that the cover of the detector should not be opened until power was disconnected (after Step 12). But Step 9 said, "After removing the cover, push the red button to discharge the capacitor."

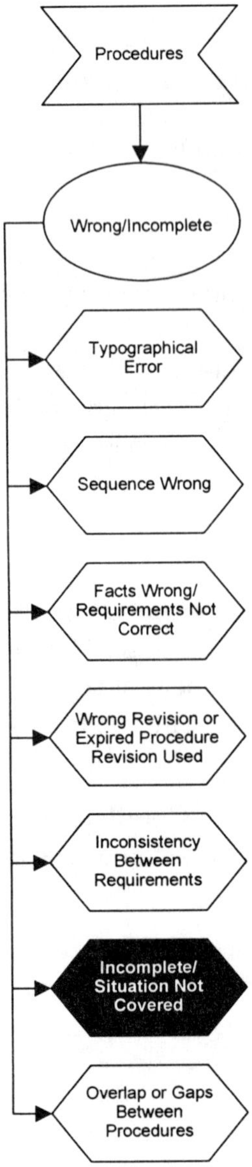

Typical Issues

Were details of the procedure incomplete? Was sufficient information presented? Did the procedure address all situations likely to occur during the completion of the procedure? Was a critical step missing?

Note: *This node addresses specific issues that are not included in a procedure. If procedures in general do not have a sufficient level of detail, consider coding under* Level of Detail LTA (Procedures, Misleading/Confusing).

Typical Recommendations

- Ensure that all modes of operation, all maintenance activities, and all special activities have written procedures
- Perform a hazard review of critical procedures to determine other accident scenarios related to errors in procedures and to determine if sufficient safeguards are provided against employees not following the written procedures

Examples

A mechanic did not correctly replace a pump. The instruction stated to "replace the pump." Numerous actions were required to replace the pump, including an electrical lockout, which was incorrectly performed.

A severe decomposition and release of chlorine occurred when the operator failed to check the strength of caustic in the neutralizer. The procedure did not include an instruction for this step, although most operators did perform this check.

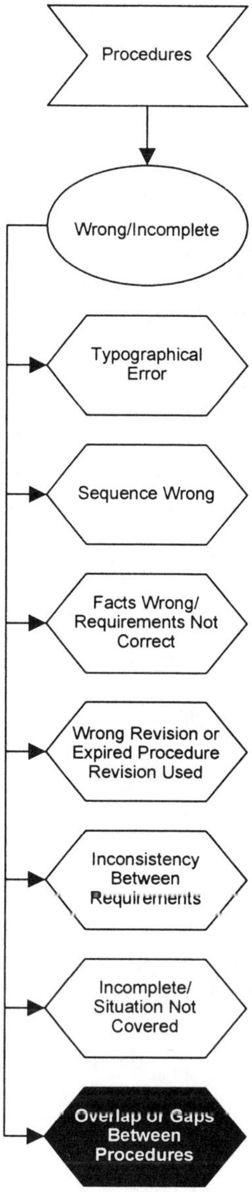

Typical Issues

Are there gaps between procedures that are used in sequence? Do multiple procedures cover the same task?

Typical Recommendations

- Develop a procedure development plan to allocate tasks between procedures
- Review procedures to determine overlaps between them
- Perform a walkthrough of the procedures to identify overlap or gaps between them

Examples

An operator started up the plant air system using the startup procedure. He then checked the normal operations procedure and it also contained a section on starting up the system.

The operator started the cooling water system using procedure CW-N-01, Normal Cooling Water System Startup. He then began the startup of three feed pumps using procedure FP-N-01, Startup of the Feed System. Gaps existed between these two procedures. Key steps were missing that were supposed to be performed after startup of the cooling water system and before startup of the feed system.

A booster pump on a pipeline was not included in the maintenance or operations procedures. The divisions upstream and downstream of the pump each thought the pump was the responsibility of the other division.

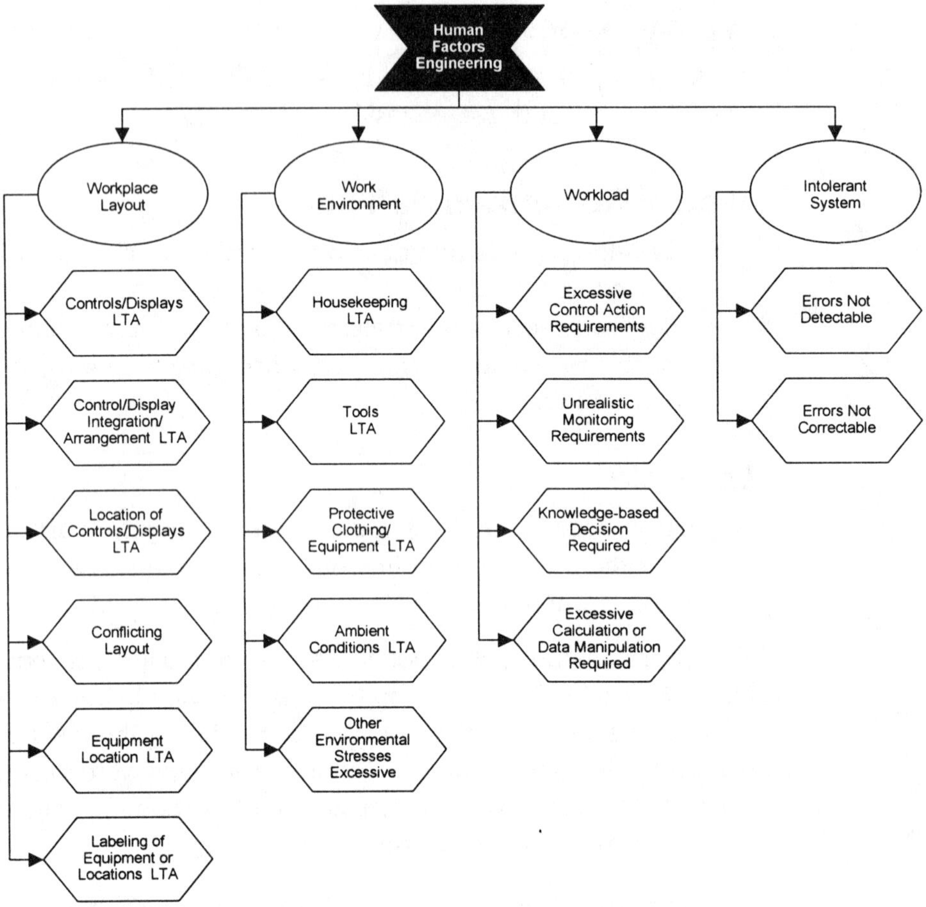

Typical Issues

Were the capabilities and limitations of humans considered in the design, development, production, and control of systems? Is the layout of the workplace adequate? Is the work environment excessively noisy, hot, or cold? Does the task impose an excessive physical or mental workload? Can the system tolerate faults?

Typical Recommendations

- Locate related controls and indications together
- Provide employees with adequate personal protective clothing such as hearing protection, gloves, and safety glasses. Ensure that they are available in different sizes to ensure a comfortable fit
- Reduce the complexity of control systems
- Provide feedback to the operator so that he/she can tell if actions are performed correctly

Examples

An operator, assigned the responsibility of monitoring a computer screen for an entire 8-hour shift, failed to detect an important signal.

An operator failed to control the flow rate in a process because the flow rate meter could not be seen from the location where the flow was controlled.

An operator inadvertently switched on the wrong pump because all three pumps switches looked the same and were not labeled.

An operator was supposed to open cartons of materials. It was difficult to obtain utility knives from the warehouse (they never seemed to have them in stock), so the operator often used a screwdriver to open the packages. As a result, some of the items were scratched by the tip of the screwdriver.

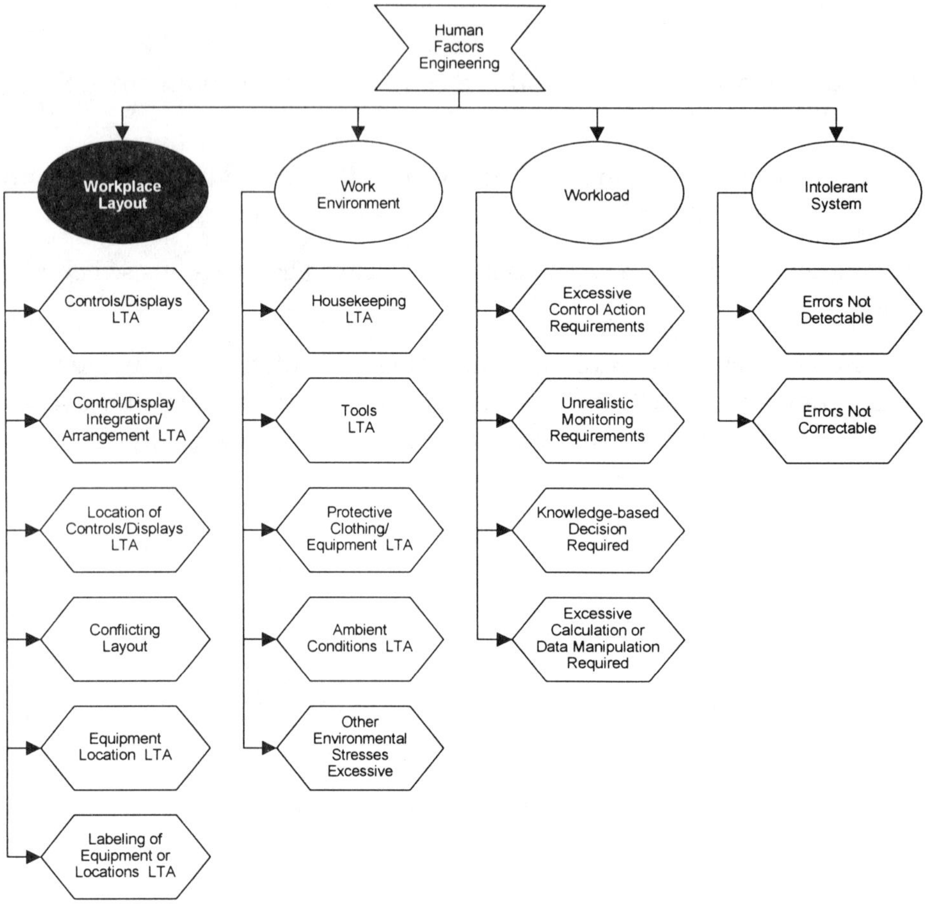

Typical Issues

Did inadequate controls or displays contribute to the error? Was poor integration of controls and displays a factor? Did differences in equipment between different processes or areas contribute to the problem? Did poor arrangement or placement of equipment contribute to the event? Was there a failure to appropriately and clearly label all controls, displays, and other equipment?

Typical Recommendations

- Ensure that operators are provided with sufficient information to control the process
- Locate related controls and indications together
- Follow expected norms in labeling and layout of controls and indications (e.g., left to right, top to bottom progression, consistent color codes)

Examples

In one processing plant, two units performed the same function. Each unit had a separate control room. The control rooms were identical except that they were mirror images of one another. An operator, normally assigned to the first unit, caused a serious process upset when he was assigned to work in the second unit.

The controller for an automatic valve was located on the front side of a vertical panel. The flow indication for the line was on the back side of the panel. A mirror was installed so the operator could see the flow indication while adjusting the valve position. However, the reversed image in the mirror caused problems in setting the correct valve position.

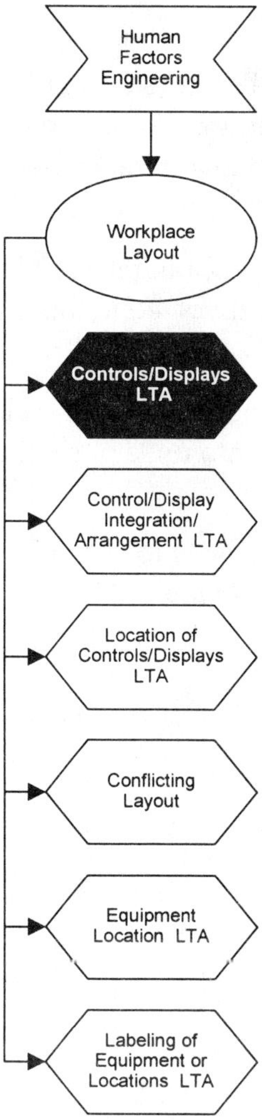

Typical Issues

Did inadequate equipment controls or control systems (e.g., push-buttons, rotary controls, J-handles, key-operated controls, thumbwheels, switches, joy sticks) contribute to the occurrence? Did the control fail to provide an adequate range of control for the function it performs? Was the control inadequately protected from accidental activation? Were similar controls indistinguishable from one another? Did one switch control a number of parameters or have different functions under different conditions?

Did inadequate displays or display systems (e.g., gauges, meters, light indicators, graphic recorders, counters, video display terminals) contribute to the occurrence? Did the display fail to provide all information about system status and parameter values needed to meet task requirements? Did the configuration of the display make information difficult to see or to interpret? Was it necessary for the user to convert information presented by the display prior to using it? Did unnecessary or redundant information contribute to the error?

Note: *Arrangement of controls is addressed by* Control/Display Integration/ Arrangement LTA. *The location of controls is addressed by* Location of Controls/Displays LTA.

Typical Recommendations

- Configure controls such that it would be difficult to accidentally activate them
- Ensure that similar controls have distinguishable features
- Ensure that the device/display allows the necessary range of control (e.g., 0-100 GPM control dial would be inappropriate if the flow sometimes required settings as high as 110 GPM)
- Ensure that sensitivity of controls allows an operator to quickly and accurately make process changes
- Ensure that displays provide enough information about the process so that the operators can adequately control it
- Configure displays so that they are easy to read and interpret
- Ensure that similar controls have distinguishable features
- Provide direct display of the necessary parameters so that operators do not have to convert the information for it to be usable
- Display only the information that is necessary/helpful to safely and efficiently control the process
- Avoid the use of dual purpose controls. Provide one control for each parameter being controlled

Examples

The operator of a remotely driven crane inadvertently dropped the load being raised. The keys on the keypad he was using to operate the crane were very small and close together. The operator's fingers, even though they were average size, were too large to accurately press one button without inadvertently pressing the surrounding keys.

During an emergency, an operator made the event worse by increasing flow instead of stopping flow. All flow controllers in the plant were moved counterclockwise to reduce flow except for the one involved in this event. It was moved clockwise to reduce flow. This violated the standard practice at this plant.

An operator made an error in reading a meter because of the unusual scale progression. Instead of a scale with major markings divided by units of five (i.e., 5, 10, 15, 20), the scale was divided into units of six (i.e., 6, 12, 18, 24)

A digital display was used to monitor the flow rate of a system. The system responded slowly to control changes. This required the operator to write down values at various times to create a time log. A chart recorder would have been a more appropriate type of display.

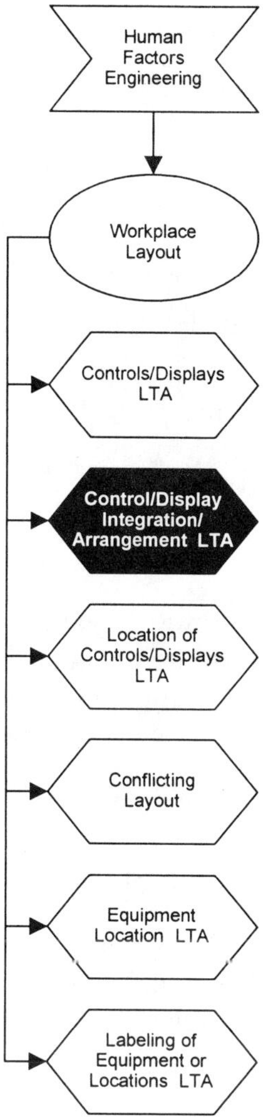

Typical Issues

Was there a failure to arrange related controls and displays of the readouts of these controls close to each other? Was a display arranged so that it was obscured during manipulation of the related control? Were control/display relationships unclear to the user? Was the response of a display to control movements inconsistent, unpredictable, or incompatible with populational stereotypes or with the user's expectations? Was there difficulty with multiple displays being operated by a single control? Is there a clear relationship between the controls and the displays? Were controls located near the displays they affected? Can the operator read the display while adjusting the control? Are control/display arrangements consistent with populational stereotypes?

Typical Recommendations

- Configure the control panel so that it is easy to locate related controls and displays
- Locate displays so that the related control can be manipulated while watching the display
- Ensure that the control and its displays are directly related to one another (i.e., if pressure is displayed, the corresponding control should directly affect pressure as opposed to another parameter, like temperature)
- Ensure that each display responds consistently with populational stereotypes when the control is manipulated (e.g., the display shows a quantitative increase when a control is turned clockwise)
- Ensure that one display is provided for every control
- Ensure that there is clear mapping between the controls and displays

Examples

The temperature control had numbers on the dial that ranged from 0 to 100. The temperature indication also ranged from 0 to 100 °C. However, setting the dial to 75 did not result in a temperature of 75 °C.

An operator set the flow rate improperly. The procedure specified the flow rate in gallons per minute. The display indicated pounds per hour.

The operator incorrectly started pump D instead of pump B. The pump controls are all identical and arranged in reverse alphabetical order from left to right like this: E D C B A This violates a stereotype that controls will be in alphabetical order from left to right.

The controls for three pumps were arranged differently than the pumps themselves.

There were three sections of lights in the room (front, middle, and back). However, the light switches were not in the same arrangement. The light switch for the back lights was located closest to the front of the room.

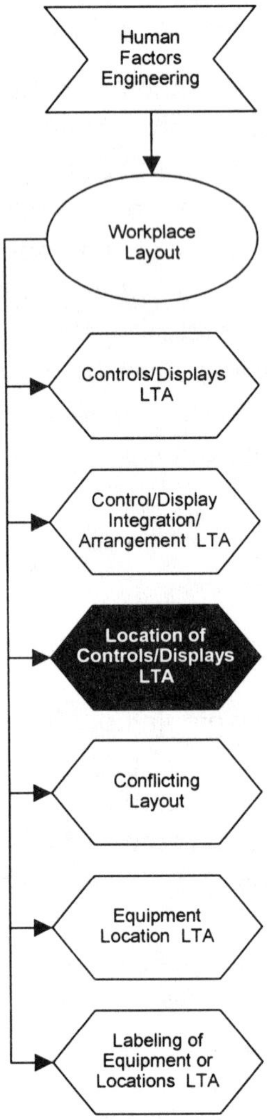

Typical Issues

Were there problems related to the location of controls or displays? Were they out of the normal work area?

Typical Recommendations

- Locate controls in convenient locations to encourage their proper use
- Locate displays in convenient locations to encourage their use
- Locate displays so that they can be read by the average person
- Locate controls so that they can be easily operated by the average person
- Locate controls so that they are not accidently bumped

Examples

A large control handle on a control panel stuck out beyond the edge of the panel when the pump was running. Someone walking past the panel accidently bumped the switch and shut down the pump. This resulted in a process upset.

The speed control for a pump was located three floors below the normal operating area. As a result, operators ignored out-of-tolerance conditions because they did not want to go up and down the three flights of stairs.

The only open space on a control panel was near the floor. As a result, a new chart recorder was installed 6 inches above the floor. To read the display, the operators had to get down on their hands and knees. Sometimes the operators just looked at the display while standing and guessed at the readings.

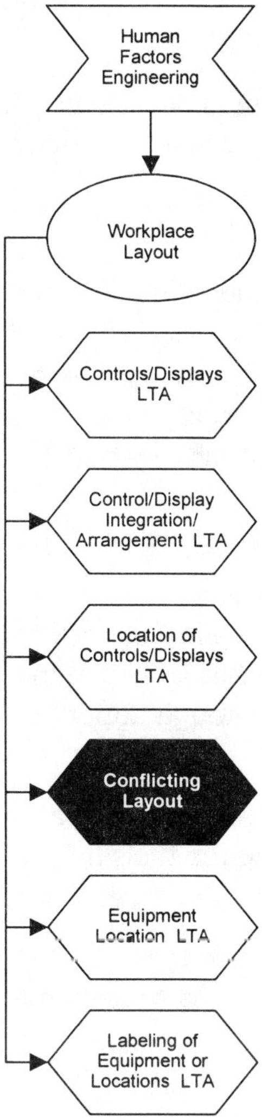

Typical Issue

Did differences in controls, displays, or other equipment between different processes or areas contribute to the event?

Typical Recommendations

- Ensure that color codes consistently have the same meaning on all control boards in the facility
- Ensure that identical units have identical control board configurations
- Label similar components in sequential order: ABC not ACB

Examples

Two computer systems, located side-by-side in the facility, were programmed using different color schemes. On the first system, the color red indicated an open valve and green indicated a closed valve. On the second system, green indicated normal and red indicated an abnormal condition. Because of the inconsistency in color coding between the two systems, an operator who normally worked on the first system allowed a tank to overflow when he was temporarily assigned to the second system. His mindset was that green indicated lack of flow.

An operator inadvertently started the wrong pump. The cooling water pumps are arranged alphabetically (A-D) from left to right. However, the control panel has the controls arranged as follows:

A C

B D

In one processing plant, two units performed the same function. Each unit had a separate control room. The control rooms were identical except that they were mirror images of one another. An operator, normally assigned to the first unit, caused a serious process upset when he was assigned to work in the second unit.

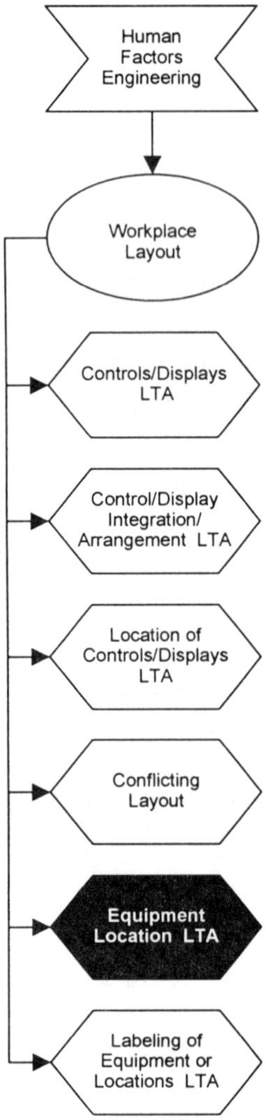

Typical Issues

Is equipment (tools, work surfaces, supplies) that personnel need to perform their jobs conveniently located? Is it accessible by workers when needed?

Typical Recommendations

- Ask workers about problems they have encountered in locating needed tools
- Locate tools and supplies so that workers will have access to them when needed. Consider back shifts and weekend access
- Review work stations to ensure that proper ergonomics are being implemented

Examples

An operator needed to make a copy of a procedure to use in the startup of a system. His printer was out of paper. The paper supply was locked in the supply room. As a result, he spent 45 minutes locating enough paper by taking it from other printers.

All tools were returned to a central tool crib each night. As a result, mechanics spent 30 minutes at the beginning of each day obtaining the tools they needed for the day and 20 minutes returning them at the end of each day.

All batch recipes were supposed to be shredded after use in the field. However, the only shredder was on the other side of the plant. As a result, many operators just threw them in the waste basket.

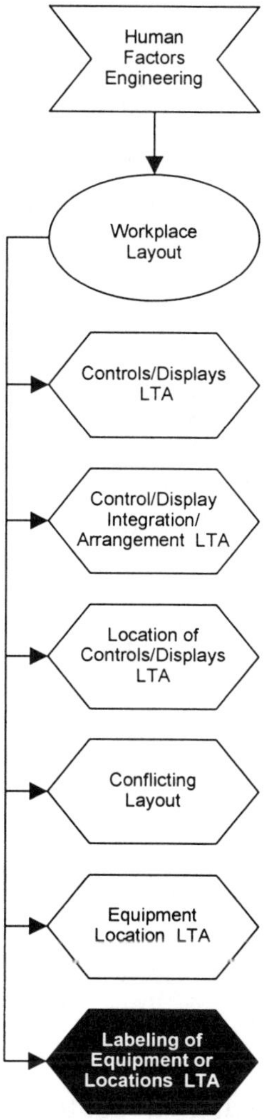

Typical Issues

Was there a failure to appropriately and clearly label all controls, displays, or other equipment items that had to be located, identified, or manipulated by the user during performance of a task? Did labeling fail to clearly identify equipment? Did labeling incorrectly identify equipment? Were labels hard to read, incorrect, or misleading?

Typical Recommendations

- Ensure that all controls and displays are labeled correctly
- Ensure that labels are made with an easy-to-read font and are color coded if necessary
- Locate all labels close to the related control/display
- Maintain labels as necessary (clean, ensure reliable adhesive, etc.)
- Ensure equipment locations or locations of materials are properly labeled
- Ensure equipment bins in the warehouse are properly labeled

Examples

An operator selected the wrong valve from a configuration of 20 valves because more than half of the valves in the group were unlabeled. The adhesive used to attach labels to the valves was not reliable in the acidic environment in which the valves were located; therefore, many of the labels had fallen to the floor. The operator tried to judge which was the correct valve using the labels that remained attached.

An operator opened the wrong valve, causing a transfer error. The label was positioned between two valves, forcing the operator to choose between them.

A row of bins in the warehouse contained different types of bolts. The labels for the bins had part numbers on them, but no equipment descriptions. As a result, some items were incorrectly restocked after being returned to the warehouse.

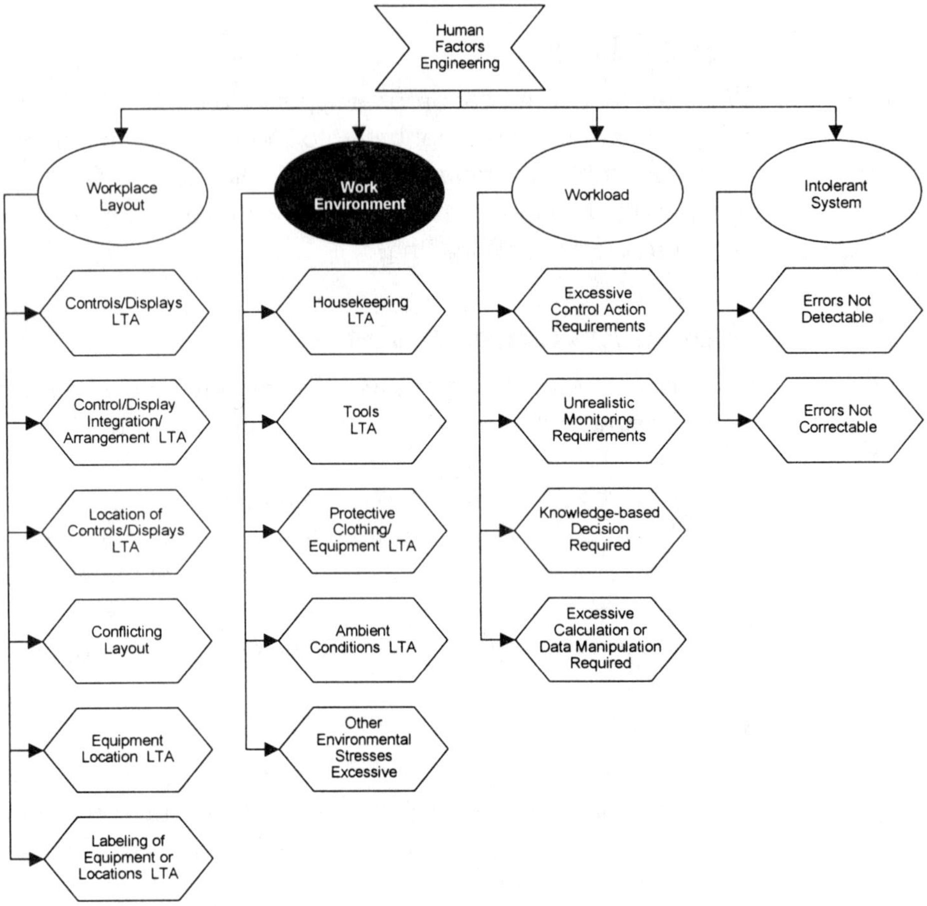

Typical Issues

Did stressors in the work environment, such as poor housekeeping, extreme heat or cold, inadequate lighting, or excessive noise, contribute to the error? Was the problem caused by difficulties associated with protective clothing? Were there other stressors present in the work area that may have contributed to the problem (e.g., vibration, movement constriction, high jeopardy or risk)? Were the right tools available to do the job?

Typical Recommendations

- Remove unused equipment and piping
- Provide employees with adequate personal protective clothing such as hearing protection, gloves, and safety glasses. Ensure that they are available in different sizes to ensure a comfortable fit
- Ensure that the right tools are available to do the job

Examples

An operator received a cut to her head when she bumped into an overhead pipe. The lighting in the area was not sufficient to detect overhead obstacles.

A step was missed during performance of a job. The operator hurried through the job because it required him to wear a respirator and work in a confined space. None of the available respirators fit comfortably.

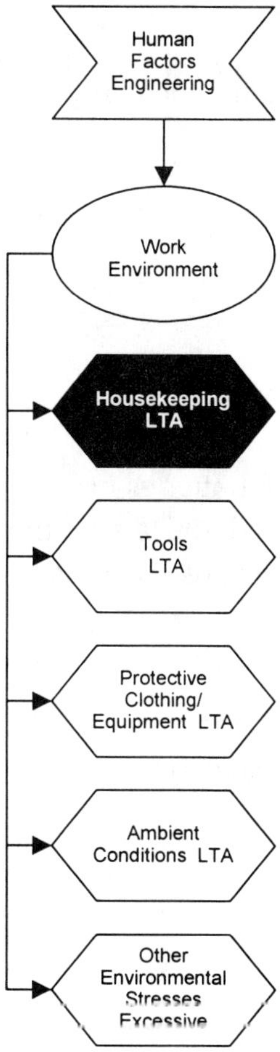

Typical Issues

Did poor housekeeping conditions contribute to the event? Was the error caused by a cluttered work environment? Was an unsafe situation created by a sloppy workplace?

Typical Recommendations

- Ensure that work areas are maintained in a clean, organized manner
- Remove (demolish) unused equipment and piping

Example

A mechanic received a puncture wound to his hand when he reached into a tool box and came into contact with an open pen knife. The tool box was full of old rags and crumpled paper; therefore, the mechanic was unable to detect the hazard.

An operator needed to check the operating records from a couple months ago. The records were stored on magnetic tape cartridges. The cartridges were labeled, but were just thrown in a drawer. As a result, it took the operator 25 minutes to locate the correct tape.

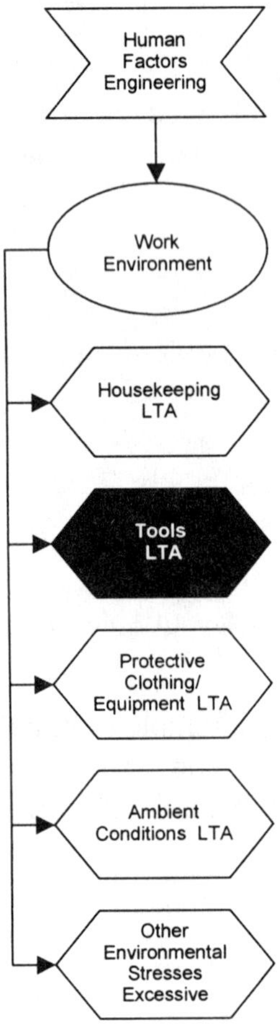

Typical Issues

Were the proper tools supplied to do the job right? Were the tools in good condition?

Typical Recommendations

- Provide the right tools to do the job right
- Ensure worn tools are repaired or replaced

Example

A maintenance helper was assigned the task of checking batteries in smoke alarms in the office areas of the plant. He was not allowed to use a voltmeter to check the condition of the 9-volt batteries (only electricians could use voltmeters). So he stuck the batteries on his tongue to see if they were still good.

A carpenter was using a hammer with a worn handle. When he was pulling out a nail, the handle broke and the carpenter injured his elbow.

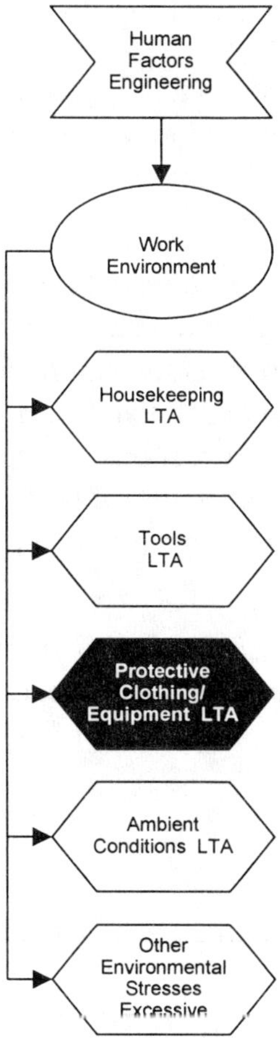

Typical Issues

Did protective clothing or equipment (e.g., plastic suit, gloves, respirator) contribute to the difficulty? Did protective clothing or equipment significantly diminish any of the senses (i.e., sight, touch, smell, hearing, or taste) necessary to perform the task? Were personnel required to wear protective clothing or equipment for an uncomfortable length of time? Were personnel required to dress in and out of protective clothing an excessive number of times?

Typical Recommendations

- Ensure that protective clothing is available in different sizes so that all employees can be properly fitted
- If several consecutive tasks require that protective clothing be worn for a long time, investigate the possibilities of using more comfortable protective clothing designs (e.g., looser or tighter fit) or protective clothing made with more comfortable material (e.g., "breathable" fabric)
- If protective clothing diminishes senses required to complete the task, investigate altering the clothing, if possible, so that personnel may perform their duties effectively

Examples

An operator wearing a full-face respirator was injured when he walked into the path of a forklift. The respirator reduced his peripheral vision; therefore, he did not see the forklift coming from his left side.

An operator using an overhead crane allowed the load to collide with operating equipment. The protective gloves he was wearing prevented him from accurately manipulating the crane's controls.

An operator splashed some alkaline catalyst onto his hands, causing a severe chemical burn, while manually loading the catalyst into a vessel. The operator was wearing gloves, but the gloves were not chemically resistant.

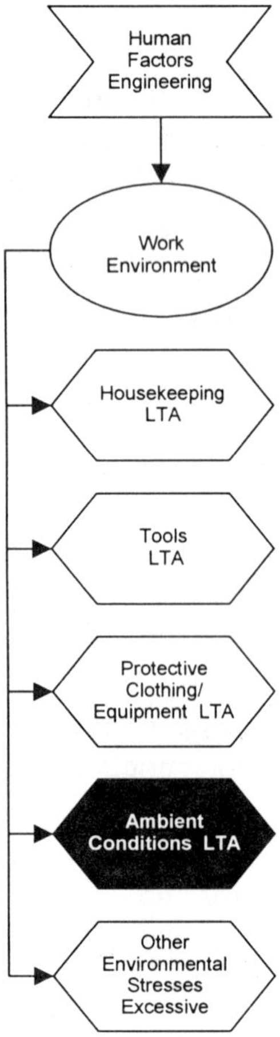

Typical Issues

Was the event caused by excessive exposure of personnel to a hot or cold environment? Was poor ventilation (i.e., poor air quality or inadequate air velocity) a contributor to the event? Was the effect of rain, snow, etc., a factor?

Was the event caused because illumination levels were not sufficient for task performance? Did the level of illumination vary greatly over a given work station? Was the error caused by failure to provide supplemental lighting for personnel performing specialized visual tasks in areas in which fixed illumination was not adequate? Was there shadowing of labels, instructions, or other written information? Was there a problem with glare or reflection? If the event occurred during an emergency situation, such as loss of power, was emergency lighting inadequate?

Was the event caused by diminished human performance caused by excessive noise? Were personnel unable to hear auditory signals or alarms because of excessive background noise? Did auditory distraction, irritation, or fatigue of personnel result from excessive noise?

Typical Recommendations

- Ensure that indoor work areas are adequately ventilated and heated/cooled
- Allow personnel to take frequent breaks if they are required to work in an uncontrolled, uncomfortable climate for extended periods of time
- Consider the need for roofing or walls over work areas for which protection from wind and precipitation reduces the hazards of operation and maintenance
- Solicit comments from employees regarding work station lighting. Address any comments received
- Provide nonglare screens for computer monitors
- Conduct an emergency drill at night and use emergency lighting. Solicit employee feedback to determine whether or not the lighting is adequate for emergency operations/evacuation
- Install additional equipment to diminish workplace noise when possible (e.g., mufflers, sound enclosures)
- Post danger signs in areas in which noise is in excess of 85 dB to alert employees to wear hearing protection in those areas
- Ensure that emergency alarms and the emergency public address system can be heard throughout the process area

Examples

During an extreme cold spell, a mechanic damaged an expensive piece of equipment by dropping a tool into its moving parts. Even though the mechanic was wearing gloves, his hands were so cold that he was unable to get a firm grip on the tool.

A serious incident occurred when glare caused by improper overhead lighting prevented an operator from detecting that an important annunciator tile was illuminated.

During a loss of power event, an operator was injured while attempting to troubleshoot the emergency generator. Lighting levels from the control room to the generator were insufficient, and he ran into a forklift on his way to the generator.

A computer operator failed to respond to a system alarm because background noise from the computer's cooling fans masked the auditory alarm signal.

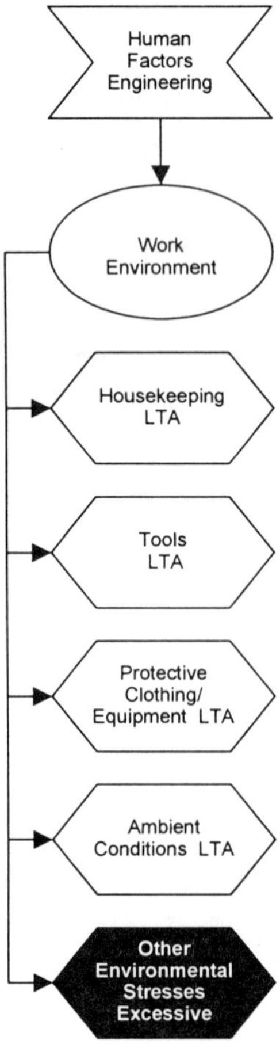

Typical Issues

Was the error a result of environmental stressors other than poor housekeeping, inadequate climate control, poor lighting, a noisy work area, or problems with protective clothing? Was the worker rushed to get the job done? Was there pressure to get the job done to allow the system to be restarted? Did he/she perceive that he/she was at risk?

Typical Recommendation

- When possible, reduce certain physiological and psychological stresses such as:
  - pain or discomfort caused by seating, etc.
  - hunger or thirst
  - vibration
  - movement constriction
  - disruption of circadian rhythm
  - high-risk job
  - perceived threat (e.g., of failure or job loss)
  - monotonous, degrading, or meaningless work

Examples

A jackhammer operator was injured when he dropped his jackhammer on his foot. He had been using the tool for several hours without relief, and the constant vibration caused his hands to "fall asleep." This weakened his grip and caused him to lose control of the jackhammer.

Working in a confined space contributed to an event because personnel rushed through the job to get out of the higher-risk environment.

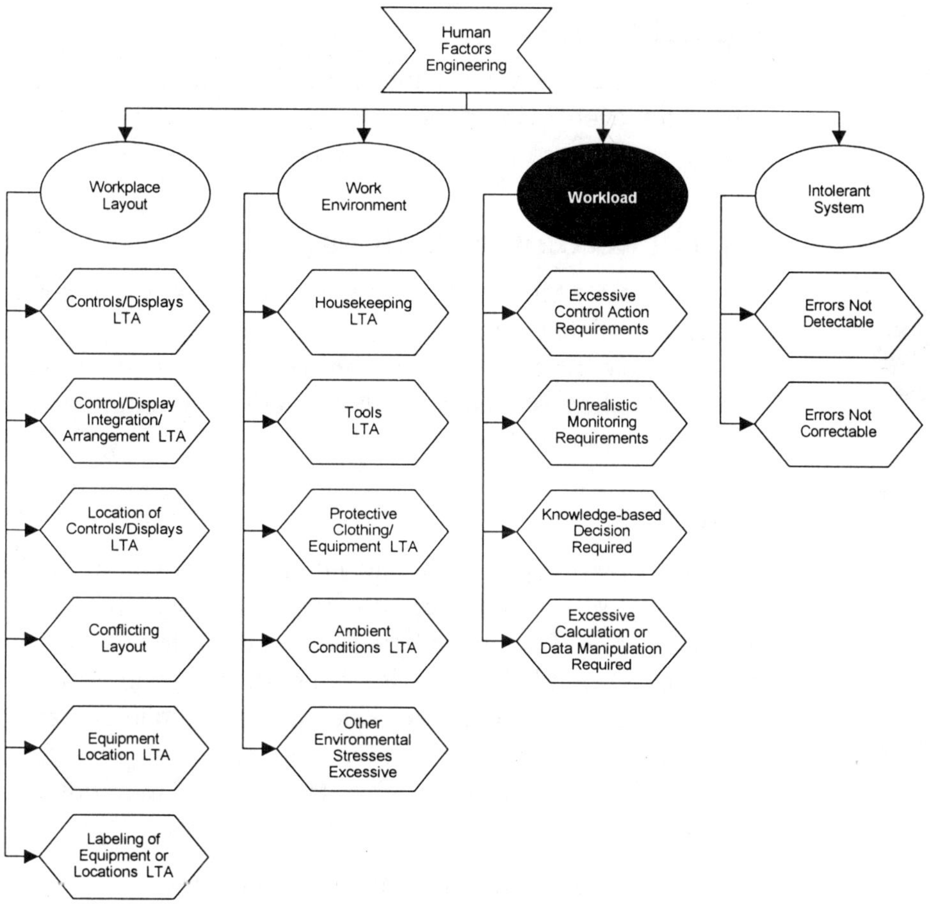

Typical Issues

Were too many tasks required for the number of available operators? Was the error caused by a situation or system being complex and requiring a decision based on specific knowledge for a successful outcome? Were system controls so complex that they contributed to user error? Did the system impose unrealistic monitoring or mental processing requirements?

Typical Recommendations

- Provide tools to make decision making easier and to reduce the chances of human error
- Reduce the complexity of control systems
- Do not place workers in situations requiring extended, uneventful vigilance

Examples

Eight maintenance tasks were in progress at the same time. The control room operator had to perform some steps for each of these tasks. He was to transfer the contents of tank A to tank B to support one of the maintenance tasks. While he was involved with another task, tank B overflowed.

The audible alarm on the toxic gas detector was inoperable. An operator was assigned to watch the toxic gas meters for an entire 8-hour shift to detect a toxic gas release. The operator failed to notice a release when it occurred.

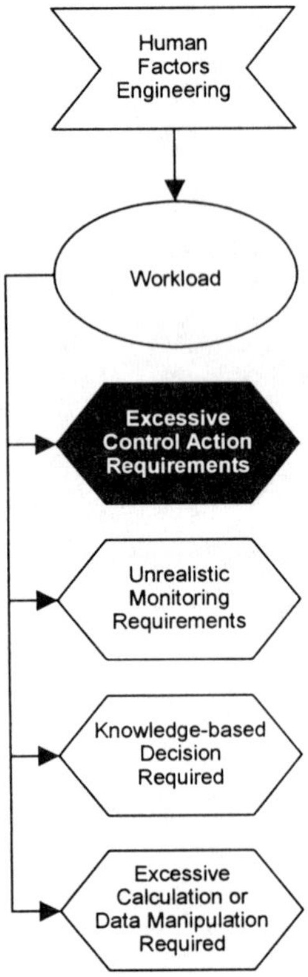

Typical Issues

Were the system or equipment controls so complex that they contributed to or caused the event? Could the system have been designed with simpler controls so that the chance of error was reduced?

Typical Recommendations

- Automate the system so that an employee is not required to constantly manipulate controls
- Reduce the complexity of the control system demands on the operator
- Make the system more stable to reduce the number of control adjustments required

Examples

A worker operating an automatic lift inside a glove box was required to operate two sets of hand controls simultaneously. These controls were located on the exterior of the glove box. One set of buttons, located on the left of the glove box, controlled the up/down motion of the lift. The other set, located on the right side, controlled side-to-side motion. While operating the lift, it was necessary for the operator to have one hand in a gloveport to balance the load. The load on the lift fell when the operator momentarily removed his hand from the gloveport to operate the controls.

The thickness of a sheet material needed constant monitoring and adjustment by the operator (15 to 20 times an hour). When other activities required his attention, out-of-tolerance conditions could go uncorrected for 5 or 10 minutes.

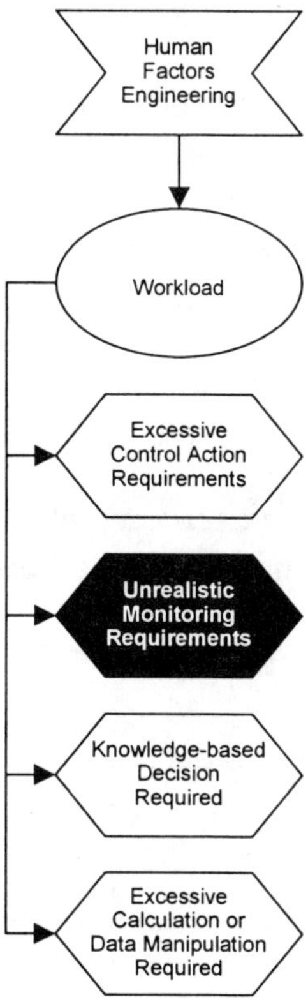

Typical Issues

Were personnel required to monitor more than three variables at once, causing overload or failure to notice important information? Could the error be attributed to loss of alertness because of the excessive length of a monitoring task?

Typical Recommendations

- Automate the system so that an employee is not required to monitor several variables simultaneously. However, provide enough employee interaction with the system to keep personnel alert
- Do not place workers in situations requiring extended, uneventful vigilance
- Ensure that staffing levels are adequate

Examples

An operator given the responsibility for temporarily monitoring the alarms for another unit allowed a tank to overflow. He acknowledged the audible level alarm for the tank, which resulted in muting of the horn. He meant to return to the problem; however, an alarm from one of the other systems sounded, and his immediate attention was required there. The tank associated with the first alarm overflowed before he was able to take appropriate action.

A radar operator was given the responsibility for monitoring a screen for blips during an entire 8-hour shift. As a result of a decrease in vigilance, the operator failed to identify an important signal.

Because of reductions in staffing levels, an operator was given the added responsibility of monitoring the operation of the flare system that serves several units, including his own. The operator can easily perform these duties during normal operations; however, during nonroutine modes of operation (e.g., startup), the operator is unable to monitor the flare system because of increasing responsibilities in his own unit. Inattention to the flare system may cause the flare system to fail to function properly, allowing a release of unburned process material to the atmosphere.

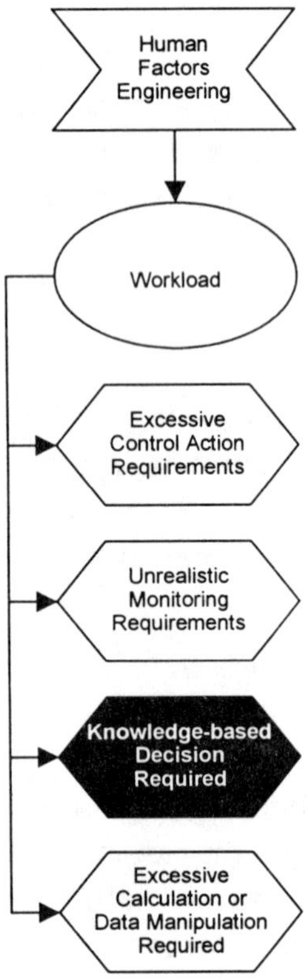

Typical Issues

Was the error caused by a situation or system being complex and requiring a decision based on specific knowledge for a successful outcome? Could better design have been reasonably expected to eliminate the error? Do personnel have to recall infrequently used information to adequately perform the task? Is it reasonable for a person to remember the information, or should it be provided on the equipment or in a procedure?

Typical Recommendations

- Modify system design to eliminate knowledge-based decision making
- Ensure that enough time is provided to complete the knowledge-based decision
- Provide tools (such as decision trees or flowcharts) to make decision making easier and to reduce chances of human error
- Provide adequate staffing to reach a knowledge-based decision

Examples

During an emergency situation, more than 80% of the annunciator tiles in the control room illuminated at once. The operators on duty were used to responding to a single alarm at a time using very specific procedures. In this situation, they did not have enough specific knowledge of how the various systems interacted; therefore, they were at a loss in determining the appropriate method of response. As a result, the operators responded to a few alarms in the wrong priority, worsening the event. In this case, knowledge of the overall system was required, and the procedures provided were useless.

The clock on a data recording unit needed to be advanced 1 hour for the switch to daylight savings time. The process for doing this was not obvious as there were no time-set buttons on the device. No procedure or directions were available for this task either. As a result, the operator tried a number of different ways before succeeding.

A line had to be flushed to clear out some contaminants. This process was only performed a few times a year. No procedure was developed for this process because it was performed so infrequently. The operator used his best judgment in performing the lineup but failed to close one valve. The backflow through this line resulted in an exothermic reaction in one of the supply tanks.

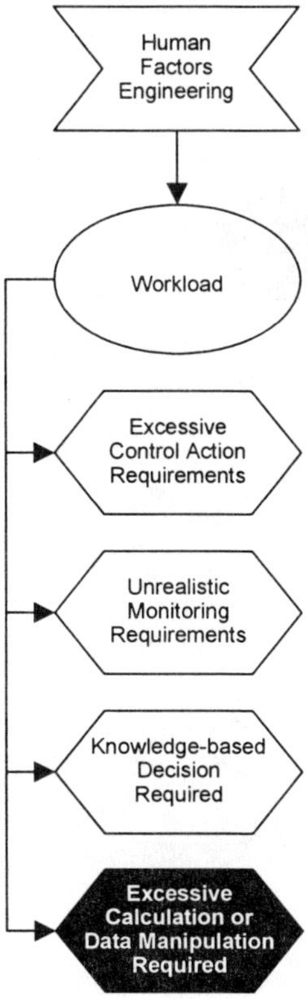

Typical Issues

Was the error due to the need for excessive mental processing by personnel? Were personnel required to work through complicated logic sequences or other written instructions? Did the task require that personnel commit extensive amounts of information to memory? Were personnel required to carry out mental arithmetic?

Typical Recommendations

- Provide workers with the information they need (e.g., procedures, calculated tables) instead of relying heavily on their mental capabilities (e.g., memory, mental calculations)
- Provide the information that workers need in the simplest form possible
- Anticipate the types of conditions workers may encounter and provide the information they will need under each of these conditions

Examples

In order to determine the amount of acid to add to a particular mixture, an operator was required to take readings from three meters and perform a mental calculation. The operator made a mental error in performing the arithmetic and added the wrong amount of acid to the tank.

An operator was attempting to determine if the present plant condition was acceptable. To do this, the operator had to determine the pressure and temperature of a vessel, then use a 60-page table to determine if the vessel had adequate subcooling by determining the saturation temperature for the pressure of the vessel. Then he compared the vessel temperature to the saturation temperature. This task could have been simplified by using a graph, a job aid, or letting the plant process computer perform the task.

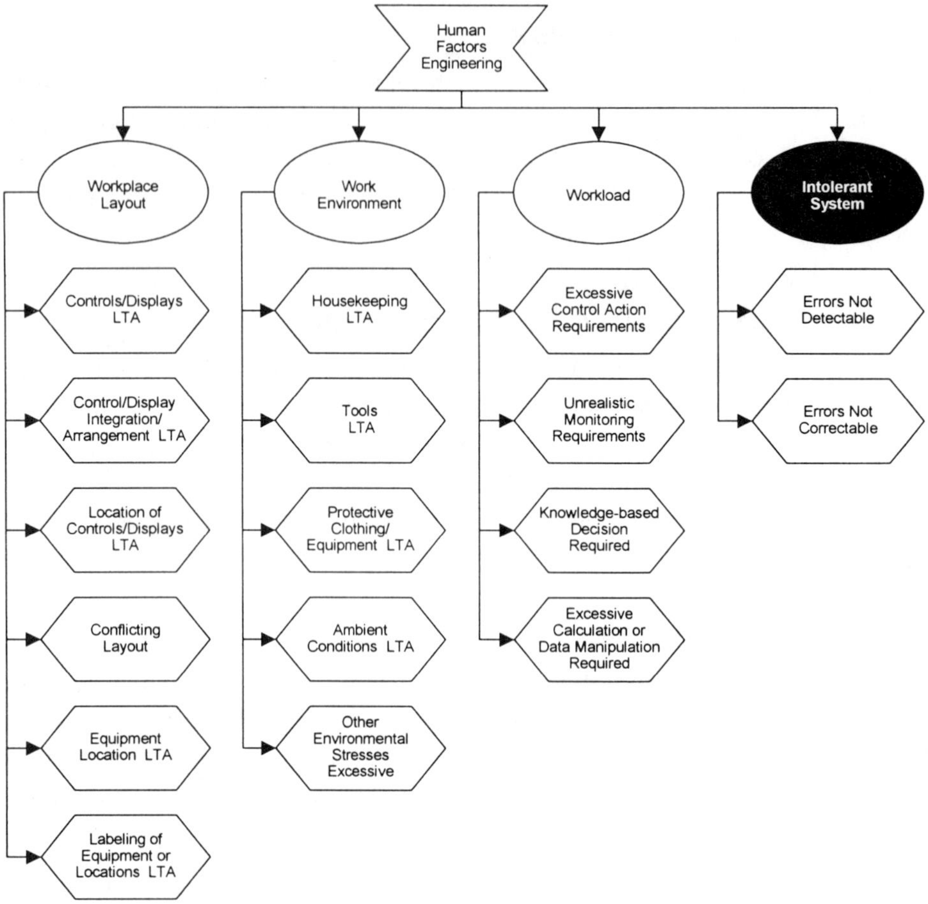

Typical Issues

Were personnel unable to detect errors (by way of alarms or instrument readings) during or after the occurrence? Was the system designed such that personnel were unable to recover from errors before a failure occurred?

Typical Recommendations

- Ensure that important safety-related equipment is adequately equipped with error-detection systems
- Provide feedback to the operator so that he/she can tell if procedure steps are performed correctly
- Design tasks and equipment to allow time to detect and correct errors for safety-critical tasks and equipment

Examples

An operator was simultaneously filling two large vessels with gasoline. While attending to one of the vessels, he allowed the other one to overflow because no level alarms or indicators were provided to let him know that the vessel was reaching its capacity.

An operator thought he closed a valve on the feed line to a tank. However, the valve stem was binding and the valve was half-open. No position indicator was provided for the valve and no flow indication was provided for the line.

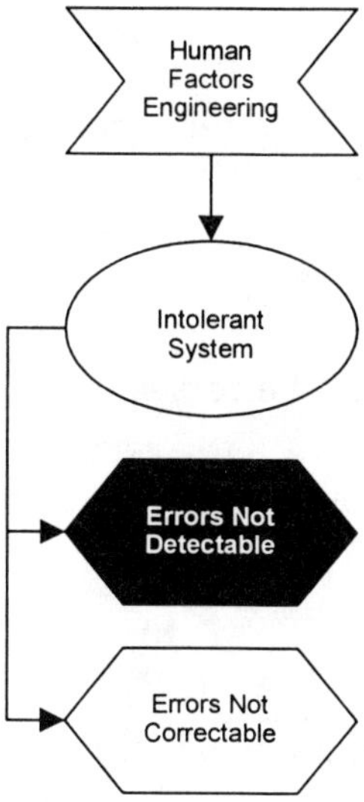

Typical Issues

Were personnel unable to detect errors (by way of alarms or instrument readings) during or after the occurrence? Did a serious error go unnoticed because no means were provided to monitor system status?

Note: *Consider dual coding with* Controls/Displays LTA (Workplace Layout).

Typical Recommendations

- Ensure that important safety-related equipment is adequately equipped with error detection systems
- Ensure that systems important to reliability and quality are equipped with error detection systems

Examples

An operator intending to stop flow to a tank accidentally turned the wrong valves. No level alarm was provided on the tank to indicate that overflow was imminent; therefore, the tank overflowed.

The alarm limits for cooling water flow were set very close to the normal values. The alarm went off frequently. The operators learned to ignore the alarm because it was part of normal operations. As a result, when cooling water flow stopped because of a failed pump, the operators did not respond.

A warehouse stock person obtained the wrong bolts for a job. The bolts were in bins that were only labeled with the part numbers; no part descriptions were included. Small parts like these were not labeled with part numbers. As a result, the stock person could not check that the materials in the bin were the ones that were supposed be there.

An operator attempted to open a block valve underneath a relief valve. The gate separated from the stem, so even though the valve appeared open (based on stem position), the gate was still closed and obstructing the pressure relief valve inlet.

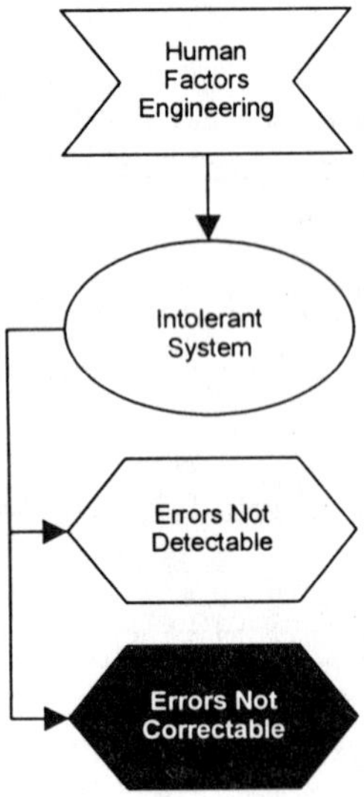

Typical Issue

Was the system designed such that personnel were unable to recover from errors before a failure occurred?

Typical Recommendations

- Design safety-related equipment so that the detected errors can be corrected before system failure occurs
- Design tasks and related procedures to allow employees time to detect and correct errors for safety-critical tasks

Examples

A computer operator started an automatic operating sequence, controlled by a distributed control system, before the valving lineups in the process area had been completed. Even though operators in the field called in to tell the operator to stop the operation, the computer was not programmed to allow interruption of the sequence. As a result, process flow was routed to waste.

A low tank alarm occurred, indicating insufficient level for the pump drawing suction from the tank. By the time the operator responded to the alarm, the pump was already damaged.

Samples were drawn of each batch prior to shipment. However, the batches were often sent out before the analysis of the samples was complete. As a result, when a sample indicated an unacceptable batch, the delivery could not be stopped before it reached the customer. The customer had to be called and asked to ship the batch back.

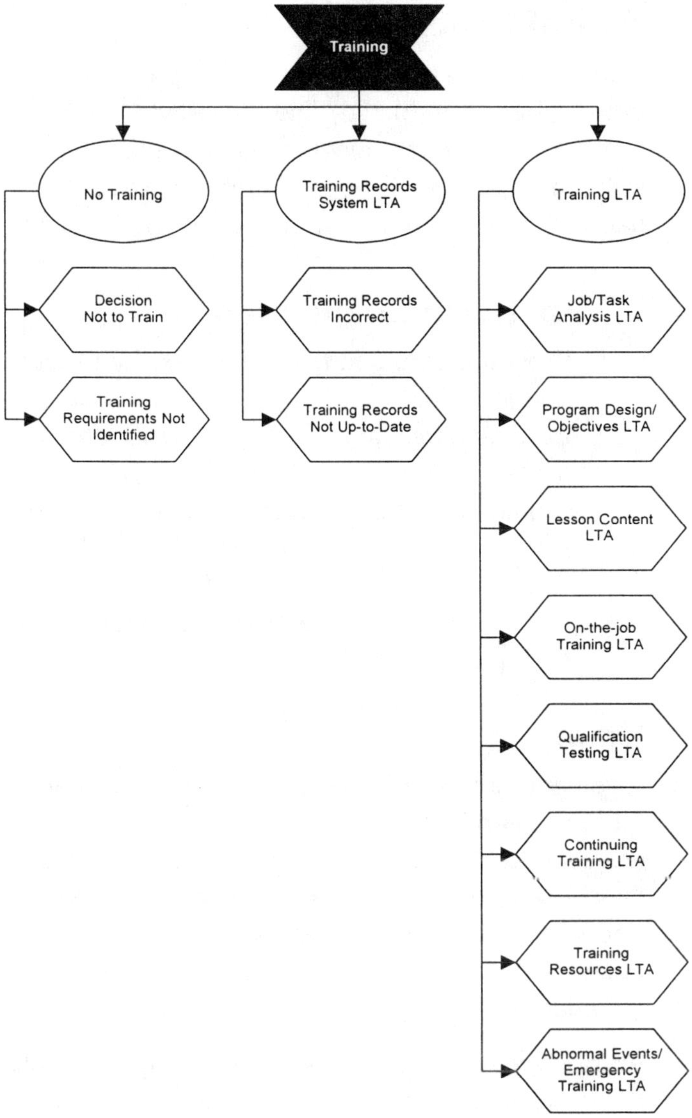

Typical Issues

Was training provided on this task? Was the training sufficient to perform the task? Did the training correspond to the actual work environment? Were training records adequate?

Typical Recommendations

- Provide training in the hazards of the process and job tasks
- Provide refresher training in appropriate areas
- Solicit comments from the trainees after they have been on the job for 3 months to determine "holes" in the training program
- Ensure that instructors are properly qualified
- Provide training on tasks critical to reliability and quality

Example

A solvent tank overflowed because the operator had not been trained on how to calculate the liquid level of a solution with a specific gravity less than water.

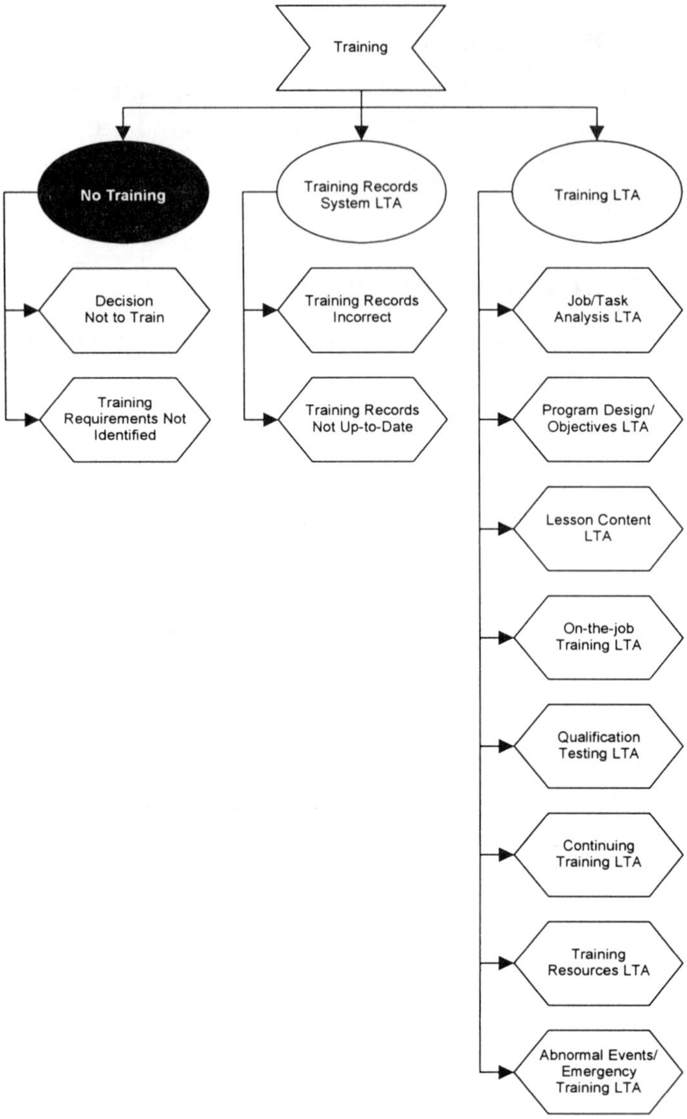

Typical Issues

Had training on the task been developed? Had training been conducted? Did the individual(s) involved in the event receive training? Had the training requirements been identified? Was a decision made to not train on the task?

Typical Recommendations

- Provide training in the hazards of the process and job tasks
- Provide refresher training in appropriate areas
- Provide a written description of the training requirements associated with a specific job title
- Provide training on tasks critical to reliability and quality

Examples

A solvent tank overflowed because the operator had not been trained on how to calculate liquid levels of tanks. Training was not required on this task because it was assumed to be a "skill of the trade." However, the operators were not experienced with solvents and solutions with specific gravities less than water.

An operator made a mistake in weighing materials. A new computerized scale had been installed a month before. Training was not provided in the use of the new scale even though it was significantly different from the mechanical type that had been used in the past.

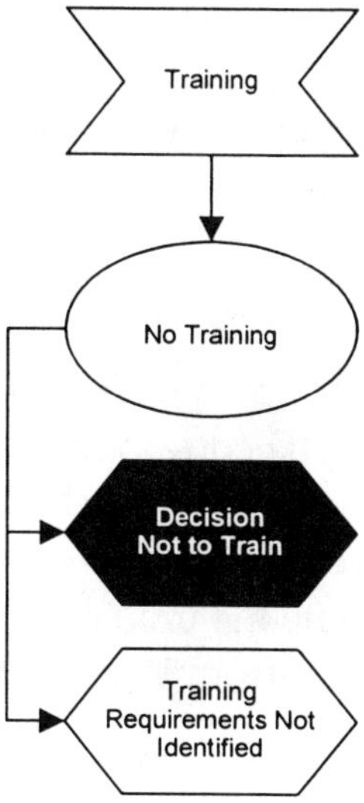

Typical Issues

Was the decision made to not provide specific training on a task? Were some employees not required to receive training? Was experience considered a substitute for training?

Typical Recommendations

- Provide training in the hazards of the process and job tasks associated with normal operations, nonroutine operations, and emergency operations
- Provide training for maintenance tasks such as inspection, testing, calibration, preventive maintenance, repair, replacement, and installation
- Provide refresher training annually for all employees in their assigned duties

Examples

A solvent tank overflowed because the operator did not know how to calculate the liquid level. The operator was not required to receive training because he had years of experience working in a similar facility. However, the previous facility did not use solvent, and the operator did not have experience with solutions with specific gravities less than water.

Management decided to only train one mechanic to repair a special digital processor used in the lab. However, while this mechanic was on vacation, the digital processor broke and another mechanic had to fix it.

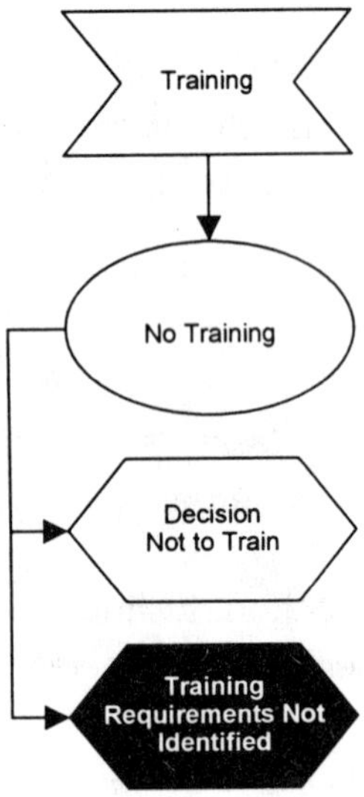

Typical Issues

Was training on the task part of the employee's training requirements? Had the necessary training been defined for the job description?

Typical Recommendations

- Identify all of the specific duties associated with each job title. Include pertinent topics associated with these duties within the corresponding training module
- Provide a written description of the training requirements associated with a job title. Require that each employee complete the training and qualification associated with his/her job title before performing specific job tasks unsupervised

Example

An operator overflowed a solvent tank because he did not know how to calculate liquid levels. The operator had transferred from a similar facility, and the training required for his present assignment had not been defined. Since the other facility did not use solvent, the operator did not have experience working with solutions with specific gravities less than water.

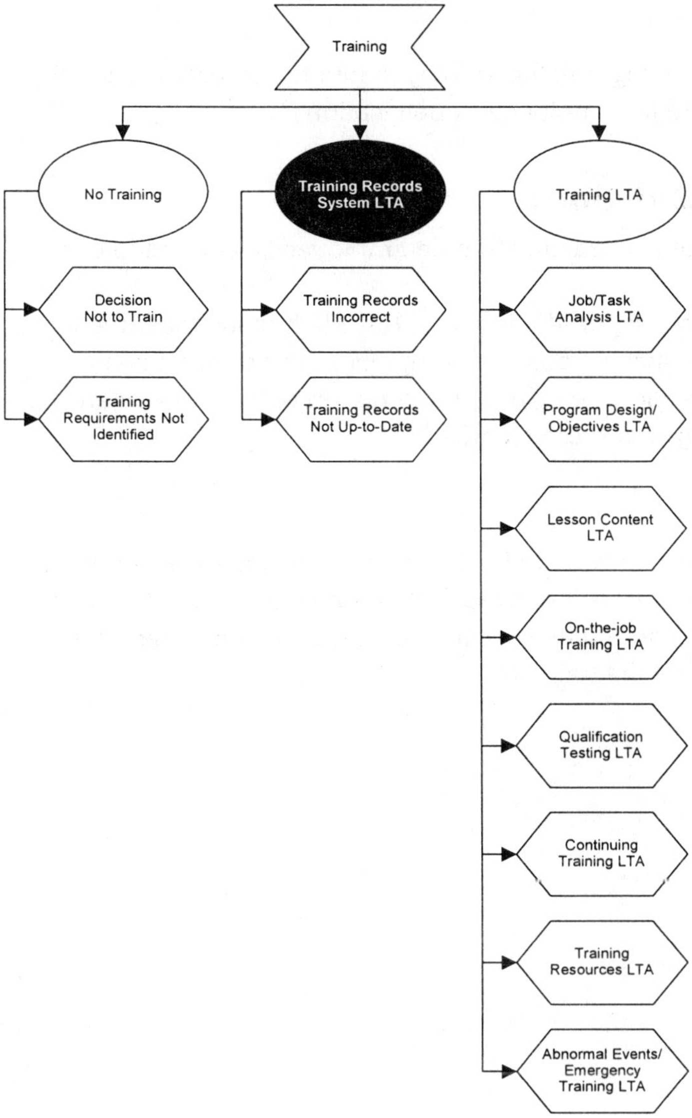

Typical Issues

Was the training record system complete and up-to-date? Did it accurately reflect the employee's training? Were the records used to determine worker selection and assignments to tasks?

Typical Recommendations

- Document the training that an individual is required to receive prior to qualification and to maintain qualification
- Ensure that individuals are assigned responsibilities for maintenance of training records

Example

A tank overflowed because the operator had not received training on how to calculate liquid levels. The training records were not routinely updated; therefore, the worker who was assigned to the job was assumed to understand this task.

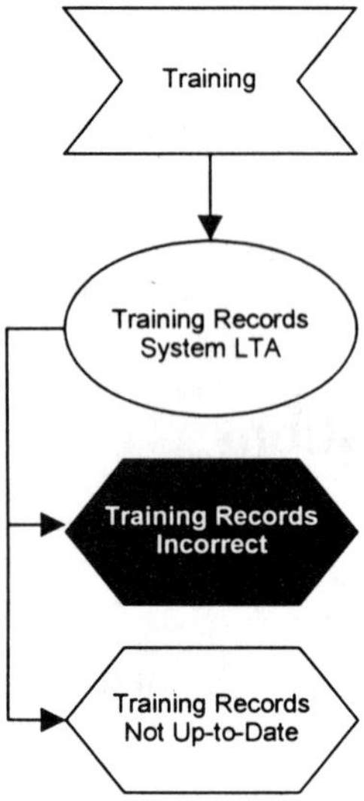

Typical Issues

Did the records show training that the employee had not received? Did the records correctly indicate the employee's qualifications?

Typical Recommendations

- Document the required training that an employee is required to complete annually
- Document all in-house, on-the-job, and outside training that an employee completes. Include dates of completion, test scores, instructor comments, certifications, etc., and a description of how competency is ascertained, along with these records

Example

An operator overflowed a solvent tank. He had been given the assignment to fill the tank because his records indicated that he had been trained on calculating liquid levels of solutions with specific gravities less than water. The operator had not received the training.

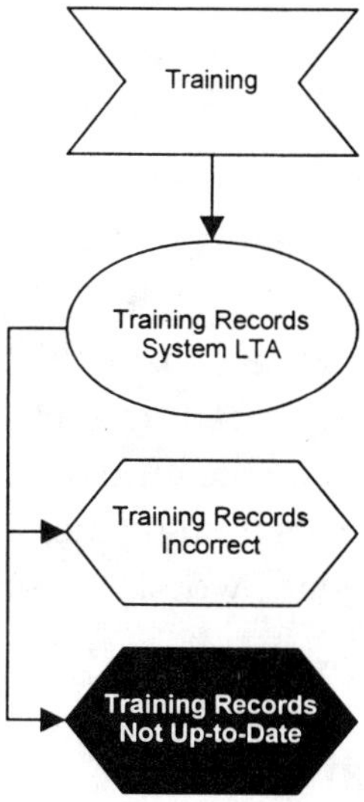

Typical Issues

Did the training records show the employee's current status for job qualification? Was the qualification expired but not reflected in the training records?

Typical Recommendation

- Establish a training records management system that assigns certain individuals the responsibility for:
  - notifying records management personnel of employee training completion dates
  - recording training completion dates
  - forwarding materials to records management personnel that verify employee understanding of the training
  - alerting employees and supervisors of upcoming training requirements
  - scheduling employees and instructors for specific training modules

Example

An operator overflowed a solvent tank because he had not received training on calculating liquid levels for solvent solutions. He had been qualified before this training was made part of the qualifications. The training records still showed him as qualified because they did not reflect the new requirements.

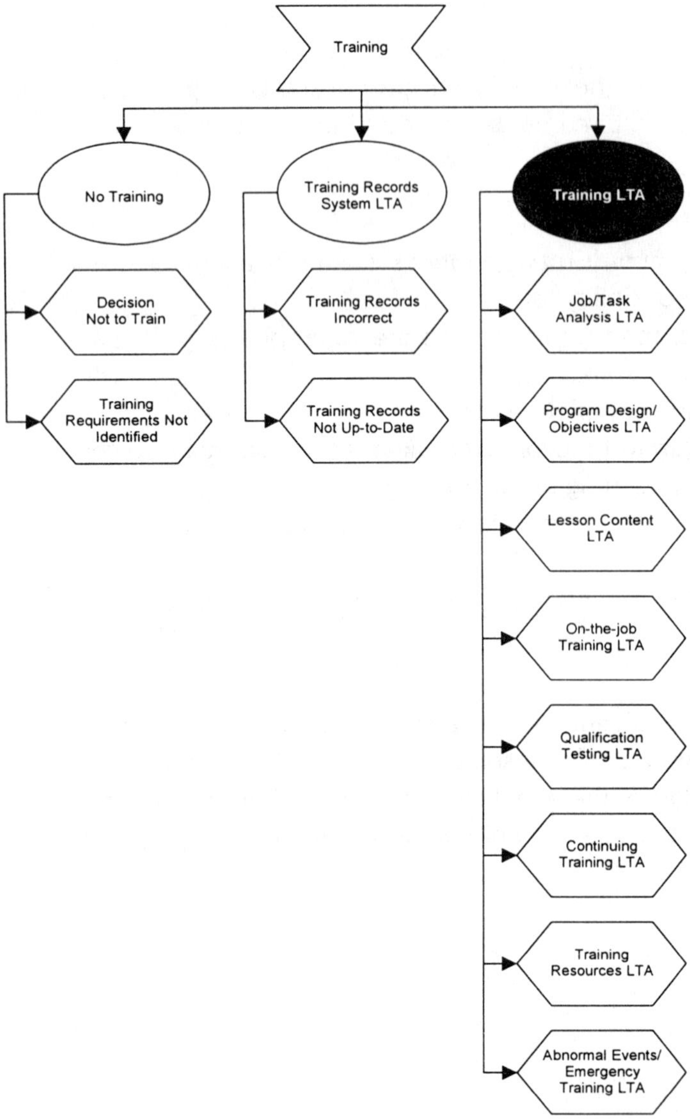

Typical Issues

Were job/task analyses adequate? Were the program design and objectives complete? Did the training organization have adequate instructors and facilities? Is refresher training performed? Does testing adequately measure the employee's ability to perform the task? Does training include normal and abnormal/emergency working conditions?

Typical Recommendations

- Perform job/task analyses for routine jobs/tasks
- Solicit comments from the trainees after they have been on the job for 3 months to determine "holes" in the training program
- Ensure that on-the-job training consists of "doing" rather than just "watching"
- Provide refresher training for nonroutine tasks
- Ensure that instructors are properly qualified

Examples

A solvent tank overflowed because the operator did not know how to calculate the liquid level of solutions with specific gravities less than water. The training included instruction in how to calculate the liquid level but did not include testing to determine if the operator could perform the calculations.

An operator made a mistake in weighing material because he used the scale incorrectly. The scale he used in training was the previous model and it had key differences from the one used on the job.

A mechanic made a mistake when repairing a pressure transmitter. Some transmitters had special seals so they would work in very high humidity environments. The job/task analysis did identify that training would be needed for these different types of transmitters.

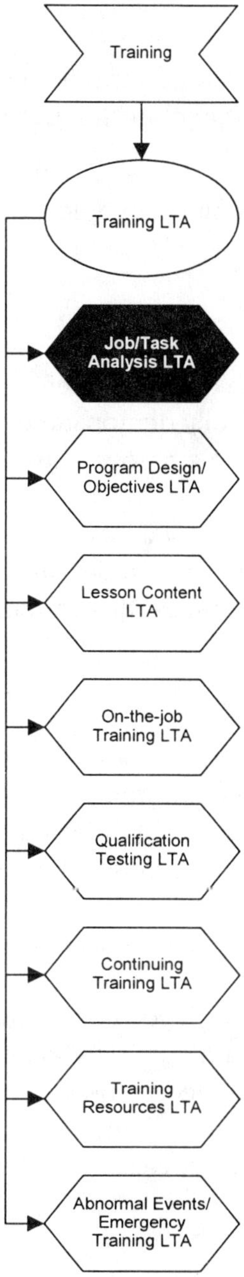

Typical Issues

Was a job/task analysis performed? Did it correctly identify the knowledge and skills necessary to complete the task? Did it correctly identify all the steps required to successfully complete the task?

Typical Recommendations

- Do not discount the value of conducting a job/task analysis for some jobs/tasks that may seem trivial or meaningless
- Include all pertinent information in the job/task analysis, including job skills required to perform the task, the sequence of task steps, and hazards of performing each task
- Conduct a walkthrough of the job/task while performing the analysis in order to trigger thoughts concerning the skills required to complete the task and the correct sequence of completing the steps

Examples

An operator made a mistake weighing material to be added to a solution. The operator had not received training on how to use the scale because the job/task analysis did not identify use of the scale as a skill for performing the job.

A technician made an error when analyzing a sample of material. The job/task analysis did not identify the need to dry the sample as part of the sample preparation.

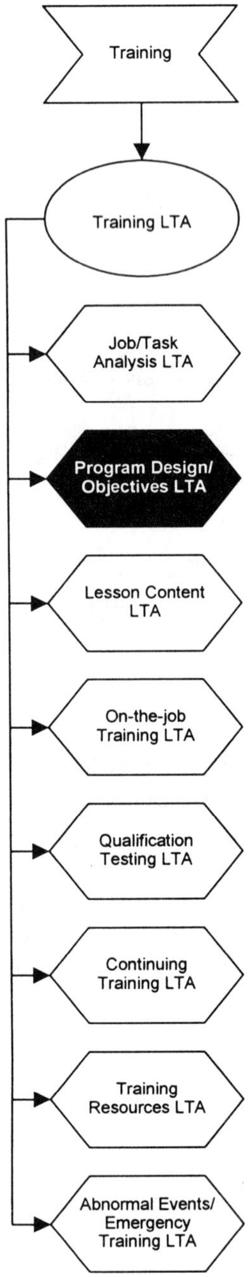

Typical Issues

Was the training program designed to equip the trainees to perform the task? Did it contain the correct amount of classroom and on-the-job instruction?

Did the objectives satisfy the needs identified in the task analysis? Did the objectives cover all the requirements necessary to successfully complete the task? Were the objectives written at the correct cognitive level?

Typical Recommendations

- Provide employees with classroom and on-the-job training. After completion of the training, have the trainee physically demonstrate all tasks (without receiving direction) to ensure that the employee has received an adequate amount of training
- After completion of a training module, have trainees evaluate the program design. Solicit comments to improve the program design
- Establish an overall training management system that assigns certain individuals the responsibility for:
  - analyzing training needs for each job title
  - establishing training criteria for each job title
  - designing curricula to meet training needs
  - continually assessing and improving the training program
- Using the job/task analysis, define and document training objectives so that employees will be equipped with sufficient skills to perform their assignments successfully
- Ensure that trainees understand training objectives at the start of each new training module
- Ensure objectives are written at the correct cognitive level. For example, the objective should be written as "Use the laboratory scale to weigh a sample" rather than "Explain how a sample is weighed." The technician's job is to perform the task, not merely to explain how to do it. Knowing and doing are on two different cognitive levels

Examples

An operator made a mistake weighing material. His formal training had contained instruction about using the scale, but on-the-job training on the use of the scale had not been required.

An operator opened the wrong valve during an emergency. In training, the operator had read the procedure but had never performed the procedure in the plant or on a simulator; nor had he performed a walkthrough.

An operator made a mistake weighing material because he used the scale incorrectly. The task analysis identified that training was required on the use of the scale, but the training objectives did not include it; therefore training did not stress this skill.

An operator overfilled a tank. The training objectives for this system required the operator to list the components in the system, but did not include an objective to explain the function and operation of the control system.

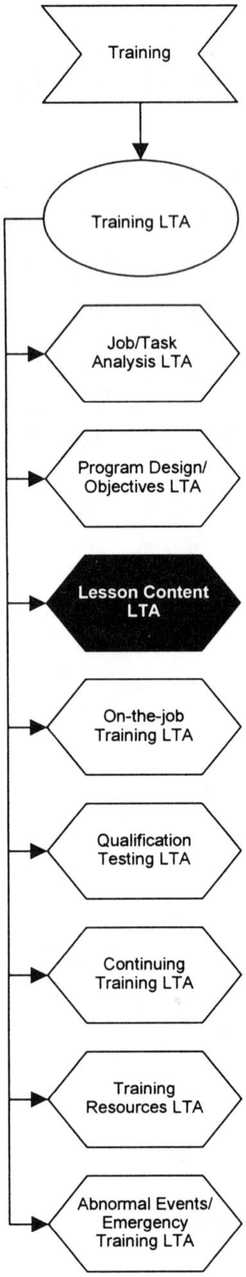

Typical Issues

Did the lesson content address all the training objectives? Did the lessons contain all of the information necessary to perform the job? Was the lesson material consistent with the current system configuration and procedures?

Note: *This node addresses the content of lessons led by training personnel or formal training away from the job (such as classroom, laboratory, or simulator training). Problems with the content of on-the-job training are addressed under the* On-the-job Training LTA *node.*

Typical Recommendations

- Ensure that the lesson content for each training module addresses all the necessary topics to guarantee a complete understanding of the required tasks
- Include workshops or demonstration techniques as part of the lesson content to provide a tangible and practical means of communication

Examples

An operator made a mistake weighing material because of incorrect use of the scale. The lesson plan did not address training on the scale, although it was in the objectives.

An operator made an error in determining the amount of material to add to a batch. The scale he used was installed 6 months before. The training he received on the system the previous month had not incorporated the new scale into the lesson content.

A clerk incorrectly entered a customized order into the computer. During training, the instructor had shown her the wrong way to perform the task.

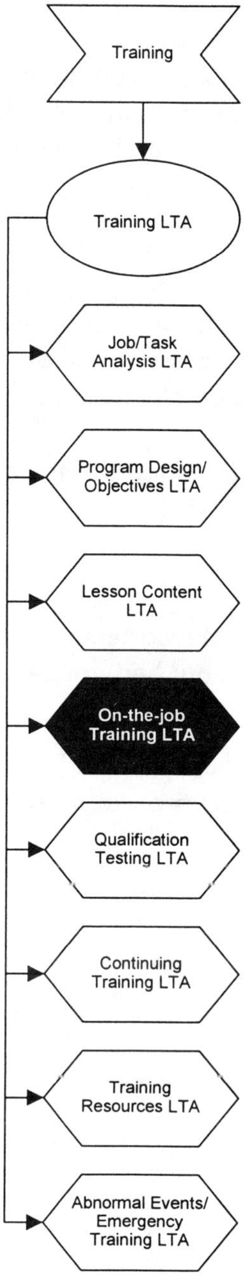

Typical Issues

Did the on-the-job training provide opportunities to learn the skills necessary to perform the job? Was there sufficient on-the-job training? Did the on-the-job training cover unique and unusual situations or equipment to avoid surprising the operator later on?

Typical Recommendations

- Ensure that on-the-job training consists of actually "doing" rather than only "watching"
- Match trainees with experienced personnel who can explain not only how to perform certain tasks, but also why certain tasks are performed
- Ensure that on-the-job training covers unique and unusual situations or equipment

Examples

An operator made a mistake weighing material because of incorrect use of the scale. He had received classroom instruction but no on-the-job experience in the use of the scale.

An operator made a mistake weighing material because of incorrect use of the scale. He had received classroom and lab instruction on the use of the scale, but the scale used in the lab was the previous model and operated somewhat differently from the one used on the job. No on-the-job training was provided.

Four furnaces were installed in a boiler house. They had each been installed at different times as the plant expanded. The control systems were similar, but had significant differences. During on-the-job training, the operator only operated two of the four furnaces. As a result, the operator accidently shut down one of the furnaces shortly after he was "qualified."

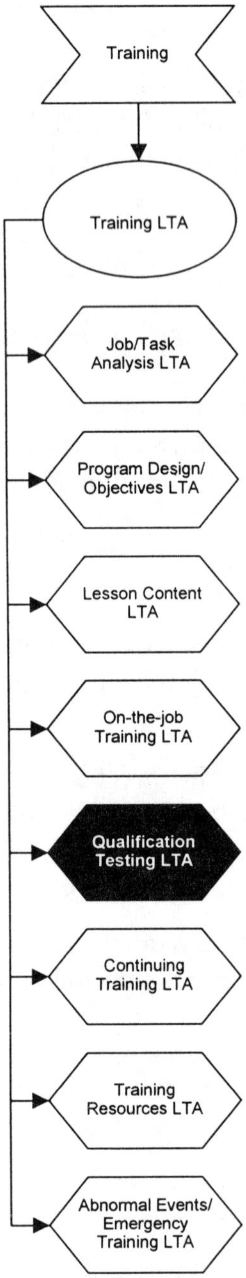

Typical Issues

Did the testing cover all of the knowledge and skills necessary to do the job? Did the testing adequately reflect the trainee's ability to perform the job? Was on-the-job demonstration part of qualification and was the demonstration thorough enough?

Typical Recommendations

- Verify that the trainee fully understood the training in some tangible manner (such as a classroom exam, physical demonstration without direction, oral exam, working with an experienced employee who is able to evaluate the trainee's performance)
- Ensure that all areas of the lesson content are verified for understanding (including both complex task skills and rudimentary skills)

Examples

An operator made a mistake weighing material because of incorrect use of a scale. He had received instruction on the use of the scale but had not been tested on his ability to use the scale.

An operator failed to close a valve in an emergency because he could not find it. Qualification testing consisted of a discussion of the procedure. A walkthrough evaluation should have been performed.

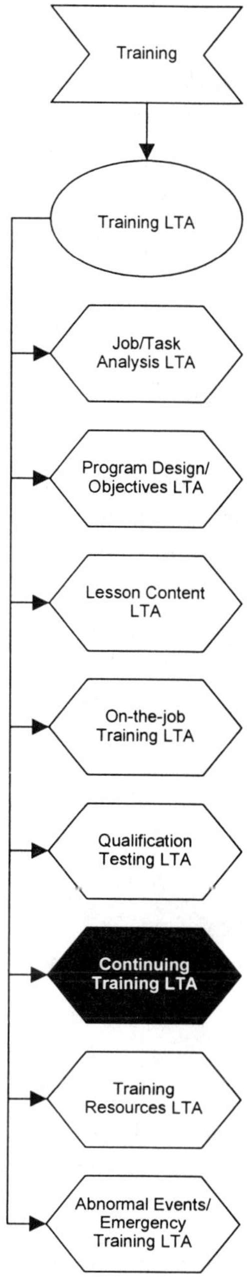

Typical Issues

Was continuing training performed to keep employees equipped to perform nonroutine tasks? Was the frequency of continuing training adequate?

Was training provided when the work methods for this task were changed? Was training provided on changes to the procedure for the task? Was training provided on new equipment used to perform the task?

Note: *Problems with refresher training on abnormal and emergency operations should also be coded under* Abnormal Events/Emergency Training LTA.

Typical Recommendations

- Provide all employees with refresher training for routine and nonroutine tasks associated with their job assignments at least on an annual basis (for operations, this would include training on startup, shutdown, troubleshooting, emergency shutdown, and safe work practices)
- Consult employees regarding the frequency of training. Should the training be conducted more often? Less often? Should refresher training content be revised?
- Provide additional training for new procedures, procedure modifications, and process modifications involving new equipment
- Ensure that the new work method training includes instructions that relate to nonroutine tasks (changes to startup, shutdown, emergency operations, etc.)
- Verify understanding of the new work method to the same degree that is required for verifying understanding of initial training (classroom exams, physical demonstration, etc.)

Examples

An operator made a mistake weighing material because of incorrect use of a scale. The scale on which he was trained had been replaced with a newer model, and no training had been provided on the new model.

A mechanic had trouble reading a graph with a logarithmic scale. The graph had been recently added to the procedure. The training department had not been notified of the change and did not identify the need to provide training on this topic.

A member of the fire team had trouble getting the foam system actuated. He received training on the system when he was hired 5 years before, but had not received any refresher training since then.

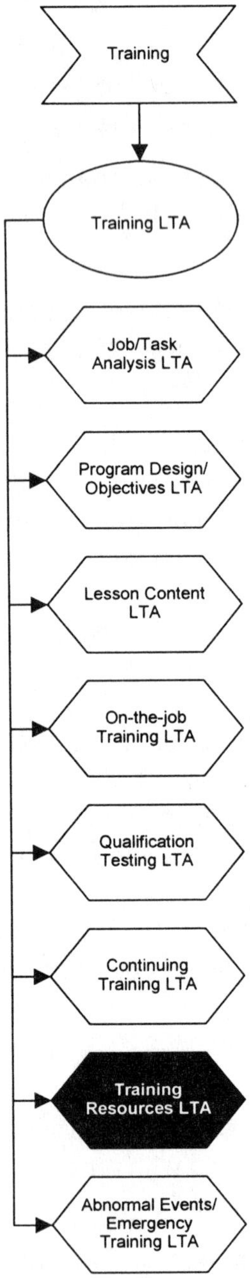

Typical Issues

Was the training equipment adequate? Were simulators or demonstration/ example components used? Was the equipment used in training the same as that used on the job? Were the instructors and other personnel providing the training adequate? Do the instructor qualifications require the instructor to be able to perform the task? Was the instructor who performed the training qualified on this task?

Typical Recommendations

- Use simulators when possible to provide personnel with hands-on experience
- If simulators are not a viable option, use models (perhaps computer models) instead. Ensure that the models are similar enough to the real equipment to avoid confusion (e.g., if a control button is actually red on the control panel, make sure it appears red on the cardboard model)
- When possible, use the same equipment in training that will be used on the job. PPE is a good example
- Ensure that proper facilities and training equipment/supplies are available for training and conducive to learning:
  - video equipment
  - overhead projectors
  - interactive workstations
  - distraction-free classrooms
- Provide guidance for determining instructor qualifications
- Review current instructor qualifications for adequacy. Address any deficiencies that are found

Examples

An operator made a mistake weighing material because of incorrect use of a scale. The scale he had used in training had key differences from the one used on the job.

A mechanic had trouble repairing a transmitter. The repair required the operator to wear gloves and a respirator. When he had practiced in training, he did not wear any protective clothing because the training department did not have any of the required protective clothing.

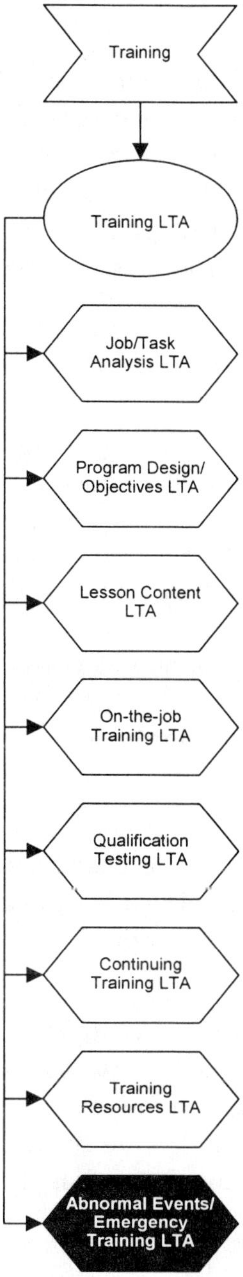

Typical Issues

Was training provided on abnormal and emergency events? Did it include all the necessary elements? Was the frequency of the training adequate?

Typical Recommendations

- Include abnormal/nonroutine job tasks as part of the initial training as well as part of the continuing training. At a minimum, the lesson content should include startup/shutdown procedures, emergency shutdown procedures, and emergency evacuation and response
- Provide refresher training for these events FREQUENTLY to give employees confidence in dealing with these stressful activities
- Establish a frequency for providing training for the abnormal/emergency events and consult employees regarding the frequency
- Ensure that the training mimics the anticipated events/emergencies as closely as practical (e.g., ensure that employees are wearing the PPE prescribed for the event when walking through the tasks)

Examples

An operator opened the wrong valve during an emergency cooling water loss. He had been trained on the emergency response during his training, but did not have to perform the task while wearing the necessary protective clothing and while responding to many alarms.

An operator opened the wrong valve during an emergency cooling water loss. He had received classroom training on the procedure, but had not performed a walkthrough or performed the procedure in the plant.

A member of the fire team had trouble getting the foam system actuated. He received training on the system when he was hired 5 years before, but had not received any refresher training since then.

During plant emergencies, management personnel were supposed to contact local authorities to coordinate evacuation of surrounding areas. During an actual event, the phone numbers that were to be called could not be found. No training had ever been provided on this task.

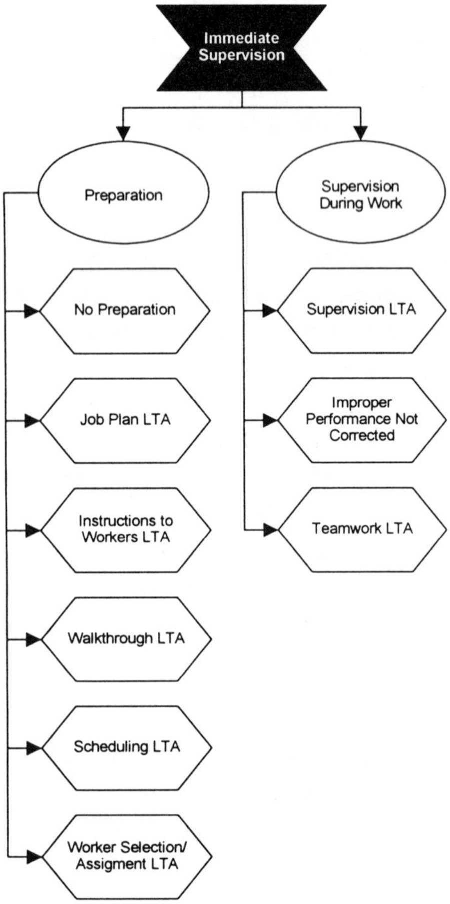

Typical Issues

Did immediate supervision fail to provide adequate preparation, job plans, or walkthroughs for a job? Were potential problems identified before the work began? Were appropriate personnel selected and scheduled for the task? Did immediate supervision fail to provide adequate support, coverage, or oversight during job performance? Did supervisors correct improper performance? Did personnel work together as a coordinated team?

Typical Recommendations

- Adopt a standard job plan format
- Distribute duties equally among similarly skilled/trained personnel
- For nonroutine jobs or jobs that require specific safety precautions, encourage supervisors to oversee the job and provide job support as necessary
- Encourage supervisors to provide more supervision to less experienced workers
- Ensure that supervisors correct improper performance

Examples

An operator failed to respond properly to an alarm because he was covering for two unit operators simultaneously. This was required because his immediate supervisor did not schedule enough control room operators to cover the shift operations.

Operators were supposed to perform plant rounds at least once per shift and generate work requests for any equipment that was inoperable or needed repairs. Often the operators skipped the rounds when it was cold or raining even though the rounds were still required. Supervisors knew what was occurring and did nothing to correct the situation.

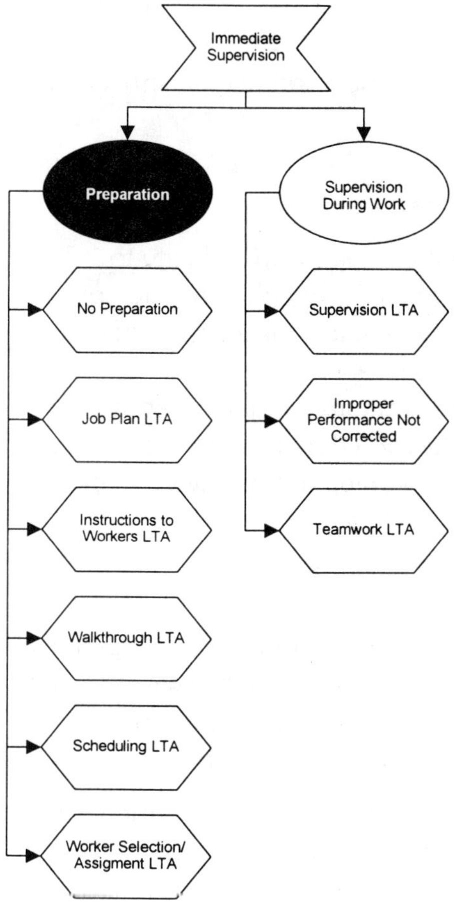

Typical Issues

Did immediate supervision fail to provide adequate preparation, job plans, or walkthroughs for a job? Were potential interruptions or special circumstances identified before the work began? Were appropriate personnel selected and scheduled for the task?

Typical Recommendations

- Ensure that supervisors understand their role in providing a job plan for subordinates
- Adopt a standard job plan format
- Distribute duties equally among similarly skilled/trained personnel
- Verify that the employee has the credentials to complete the task before assignment

Examples

Late in the shift, a first-line supervisor instructed a mechanic to repair a valve in a confined space. However, his supervisor failed to schedule anyone else to assist with the entry. To get the job done before the end of the shift, the mechanic entered the confined space alone and died.

A job required the coordinated effort of the operators, mechanics, and electricians. The electricians were the lead group on the project. The electrical supervisor failed to arrange for support from the other two groups.

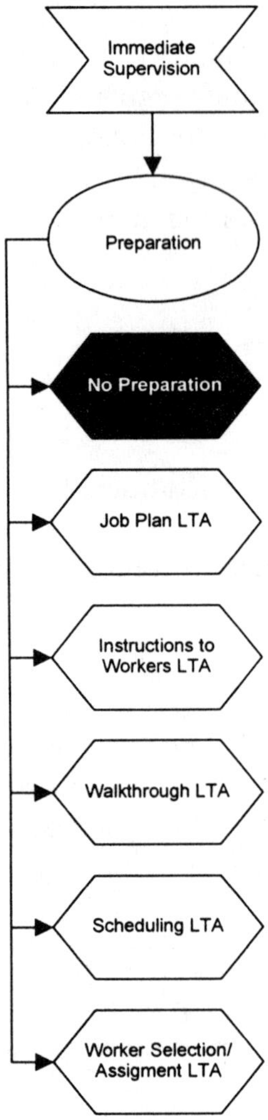

Typical Issue

Did immediate supervision fail to provide any preparation (e.g., instructions, job plan, walkthrough) for the task performed?

Typical Recommendations

- Ensure that supervisors understand that it is their responsibility to provide subordinates with instructions and/or a job plan, and to conduct walkthroughs when appropriate (to show workers the location of equipment or to discuss the proper sequence of steps, etc.)
- Provide supervisors with written job descriptions so that the above responsibilities are clearly communicated and documented
- Provide coaching to supervisors whose job preparation skills are less than adequate

Examples

An immediate supervisor sent his crew out to paint stripes in a parking lot. No instructions were given for the job. As a result, the crew used the wrong color paint to stripe the lot. In addition, the resulting parking places were not of adequate size to accommodate anything other than compact cars.

A job required the coordinated effort of the operators, mechanics, and electricians. The electricians were the lead group on the project. The electrical supervisor failed to arrange for support from the other two groups.

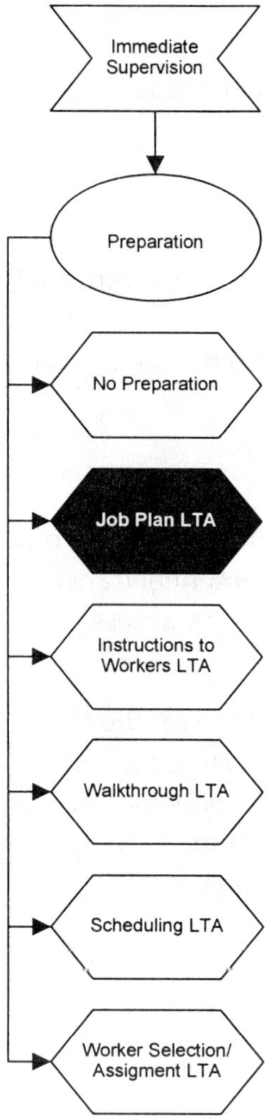

Typical Issue

Did immediate supervision provide an incorrect, incomplete, or otherwise inadequate job plan for performance of the work?

Typical Recommendations

- Establish an administrative procedure that requires all supervisors (including contract supervisors) to provide their subordinates with a job plan that includes instructions necessary for completing nonroutine job tasks
- Establish a facilitywide job plan format to ensure that all necessary information is included in the job plan

Example

A new unit was undergoing its first turnaround. During the turnaround, maintenance personnel contaminated the replacement catalyst because of handling/loading errors. The new catalyst required special handling precautions that the crew was not aware of. The turnaround plans were the same as for the old unit that was replaced and did not provide for special handling of the new catalyst.

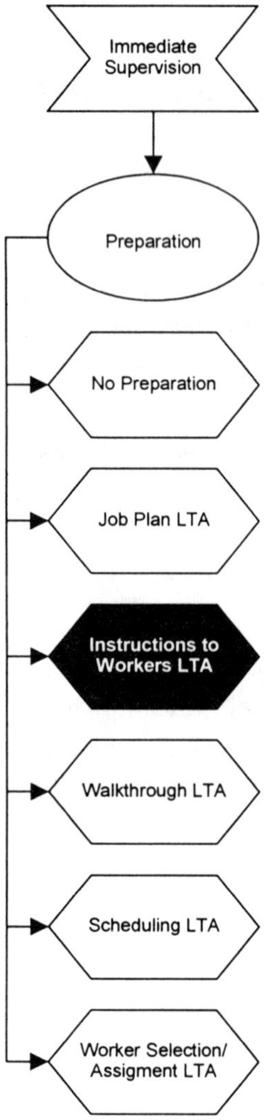

Typical Issue

Did immediate supervision provide incorrect, incomplete, or otherwise inadequate job instructions before the beginning of work?

Typical Recommendations

- Encourage a culture that is feedback oriented (i.e., repeating instructions back to the instructor to ensure understanding)
- Train supervisors on how to give instructions and how to verify that instructions are understood

Examples

An electrician was instructed to check the potential transformer on the main generator. His supervisor meant to tell him to check the potential transformer on an emergency generator. When the electrician opened the access panel on the main generator, the plant shut down.

A captain of a commercial cargo plane ordered his copilot to set the throttles to the full position. The copilot thought a plane was still taxiing on their runway and that they should not take off. But, the captain was the captain, and the copilot felt he should follow his orders without question. As a result, their plane hit another plane and more than 500 people died.

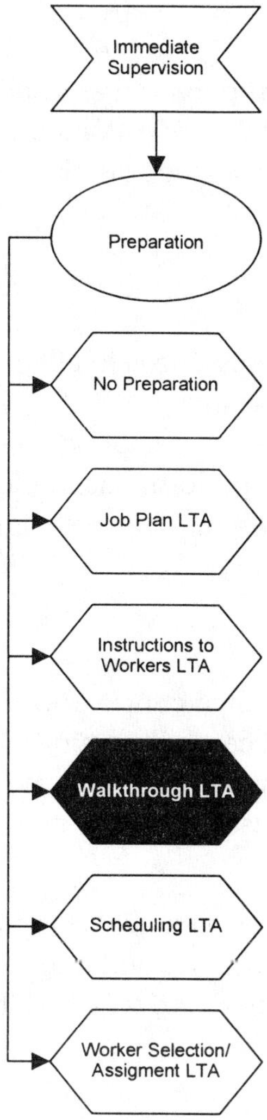

Typical Issue

Did immediate supervision fail to perform an adequate walkthrough (show workers the location of equipment, discuss operation of the equipment and the proper sequence of steps, etc.) with the workers before they started their job?

Typical Recommendations

- Encourage supervisors to show workers the location of equipment involved in the job task
- Encourage supervisors to discuss operation of the equipment and the sequence of steps involved in nonroutine job tasks

Example

A team of expert mechanics was assembled to install a special piece of equipment in a new facility. Although these were experienced mechanics, they were unfamiliar with both the facility and the specific piece of equipment. The immediate supervisor assumed that, because these mechanics were experts in their field, they did not need to be "stepped through" the job. However, the job required some special precautions, and the mechanics damaged the equipment because they were not shown the specific problem areas before starting the job.

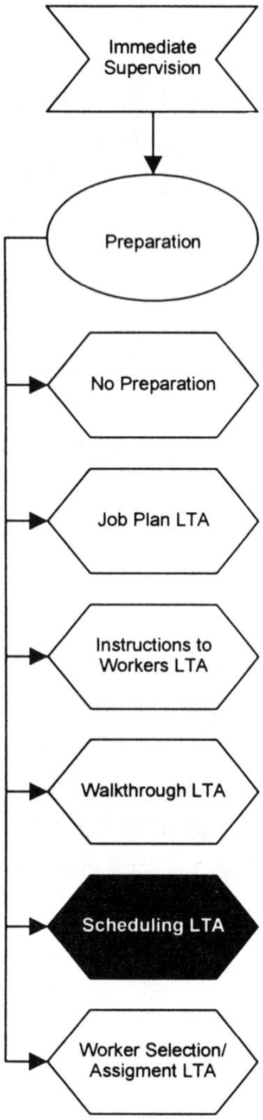

Typical Issues

Was scheduling of workers inadequate? Did immediate supervision arrange to have enough personnel available to effectively carry out the task? Were too many concurrent tasks assigned to workers? Were duties not well distributed among personnel?

Typical Recommendations

- Provide supervisors with an adequate number of employees to effectively and safely complete the tasks assigned for the shift
- Distribute duties equally among similarly skilled/trained personnel
- Consider the amount of time and concentration to perform each task. Assign individuals fewer responsibilities for tasks that require more time and concentration

Examples

As a result of inadequate planning by a first-line supervisor, a control room was staffed by one trained operator and five trainees. Because the trained operator was continuously stopped by the trainees to answer questions, he missed an important step in his own procedure. This caused a significant period of downtime in the facility.

Four mechanics and three electricians were assigned to install a new compressor. There was only enough work to keep two of the mechanics and one electrician busy. The remaining four workers just sat around and watched the others work.

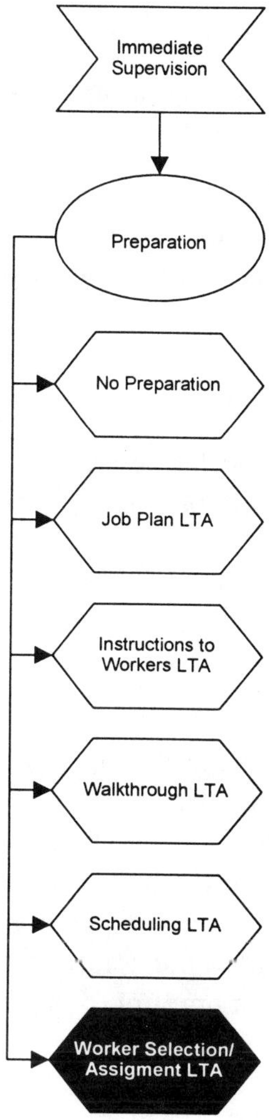

Typical Issues

Did immediate supervision fail to select capable workers to perform the job? Did workers assigned to the task have inadequate credentials? Were sufficient numbers of trained or experienced workers assigned to the task?

Note: *This node addresses the assignment of existing or qualified workers to job tasks. For example, the selection of a laborer from a preapproved pool of individuals would be covered by this node. It does NOT address the hiring or preselection processes. Employee hiring is addressed by* Employee Screening/ Hiring LTA (Administrative/Management Systems, SPACs LTA).

Typical Recommendations

- Before assigning any employee to a task, verify that the employee has the credentials to successfully complete the task
- Ensure that the individual assigned to a task matches the experience level required to effectively and safely perform the task
- Provide supervisors with the means to quickly determine if workers are qualified for a task

Example

Three technicians were assigned to a shift. Normally, at least one senior technician was assigned as the lead technician on each shift to plan and help coordinate the work. On the back shift, an older but inexperienced technician was assigned as lead technician even though he was not qualified.

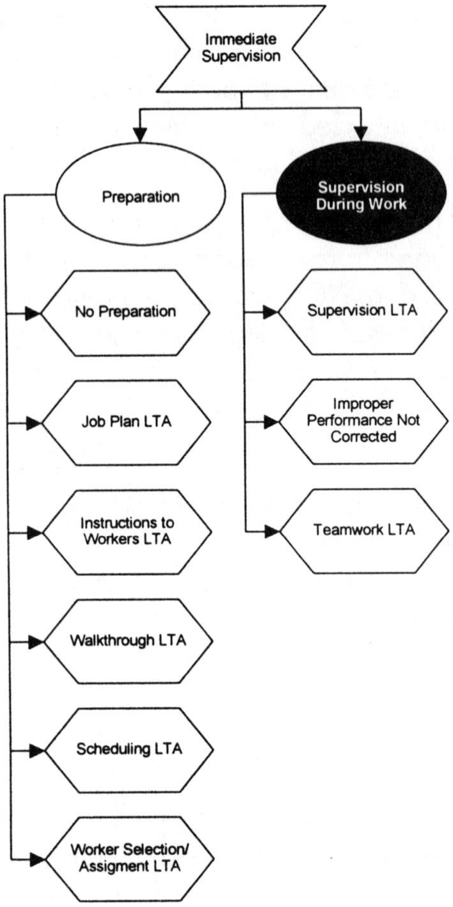

Typical Issue

Did immediate supervision fail to provide adequate support, coverage, oversight, or supervision during job performance?

Note: *The investigator must judge what level of supervision was necessary based on the importance of the job in relation to safety and production. It is not possible or practical to provide supervision on every job.*

Typical Recommendations

- For nonroutine jobs or jobs that require specific safety precautions, encourage supervisors to oversee the job and provide job support as necessary
- Encourage supervisors to provide more supervision to less experienced workers

Example

A first-line supervisor was in his office performing audits of completed procedures. He told the operator in the control room to contact him if problems arose. The operator, a newly qualified person on the job, did not want the supervisor to think that he did not know what he was doing, so he "took his best guess" when he had questions. By the time the supervisor came to the control room to check on the operator's progress, a significant amount of product had already been lost to the waste stream.

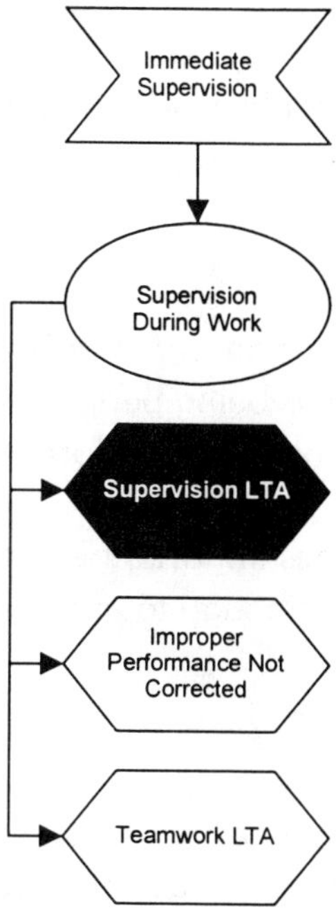

Typical Issues

Did immediate supervision provide inadequate support, coverage, or oversight during performance of the job? Was an inadequate level of supervision provided at the job site? Was contact with workers too infrequent? Did direct supervision's involvement in the task interfere with the supervisory overview role?

Typical Recommendations

- For nonroutine job tasks or for tasks that require specific safety precautions, encourage supervisors to remain at the job site to provide coverage for the entire job, or at least frequently visit the job site to provide direction as necessary
- Encourage supervisors to give their supervisory role priority over their job task support role
- Ensure that supervisors understand their responsibilities to provide more supervision to less experienced workers

Example

A mechanic was told by his immediate supervisor to "fix the leak" in a tank containing a hazardous chemical. The supervisor gave him no instructions on how to perform the task and did not provide any oversight of the work activities. Because of the mechanic's lack of understanding about the hazards associated with this job, he allowed the chemical to come into contact with his skin. This caused severe burns.

During the installation of a new computer system, the immediate supervisor of the responsible crew became so interested in installing the central control unit that he picked up a screwdriver and became involved in the work. As a result, he ignored those members of the crew who were installing the auxiliary unit. Some important checks were missed on the auxiliary unit; therefore, it failed upon startup.

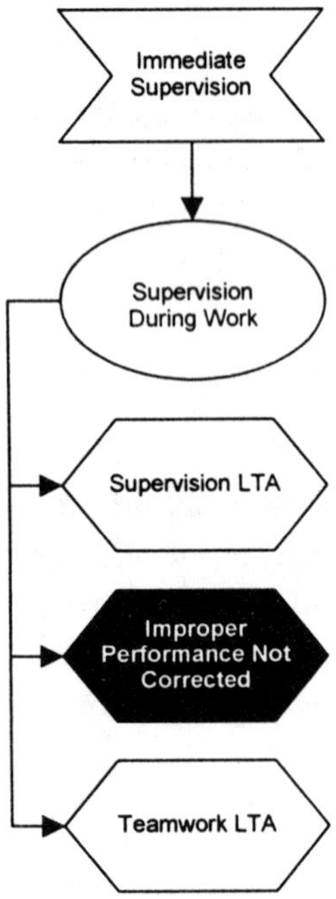

Typical Issues

Do supervisors correct improper performance when they observe it or know about it? Do they let improper performance slip "just this once"?

Typical Recommendations

- Correct the behavior when improper performance is observed or is known by supervision. If supervision knows a task is being performed incorrectly and does not correct it, workers will continue to perform the task incorrectly
- Enforce existing rules and requirements. If the rule is important enough to exist, it should be enforced. If it's not important enough to enforce, eliminate the requirement

Examples

A supervisor noticed an operator in the process area who was not wearing a hard hat or safety goggles. The supervisor was just passing through the area and did not say anything to the operator.

Operators were supposed to perform plant rounds at least once per shift and generate work requests for any equipment that was inoperable or needed repairs. Often the operators skipped the rounds when it was cold or raining even though the rounds were still required. Supervisors knew what was occurring and did nothing to correct the situation.

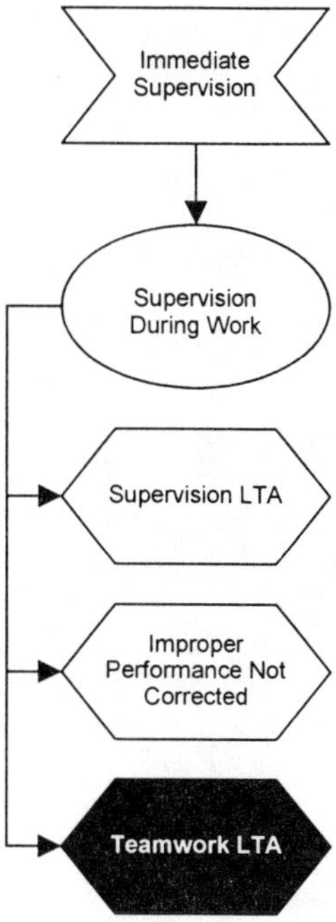

Typical Issues

Was there a lack of coordination between workers? Was a plan developed to assign responsibilities to different team members? Were there overlaps or gaps in the work that was assigned to different groups or team members? Was there a lack of communication between work groups?

Typical Recommendations

- On tasks that require coordination of work, ensure that tasks are assigned to team members and that an adequate means of communication is provided between workers
- For work that requires coordination of multiple work groups (i.e., operations, maintenance, and chemists), ensure that there are clear methods and means for exchanging information between work groups
- Coordinate tasks between different work groups. Develop a work plan prior to beginning the work

Example

Work was being performed on two different portions of a pipeline. The work performed at one booster station affected the work being performed at the receiving station. Because the work of the two groups was not coordinated, a small release of material occurred from the pipeline.

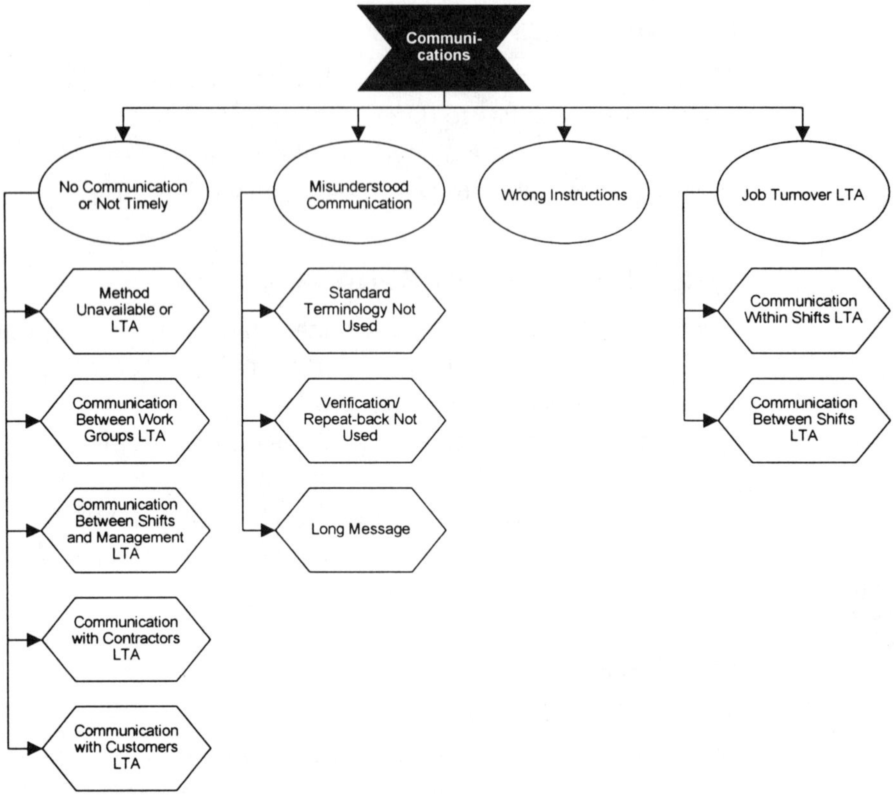

Note: *"Communications" is defined as the act of exchanging information. This node addresses many modes of communication (e.g., face-to-face, telephone, radio, short written messages, log entries). It does not address the more formal methods of communication involving written procedures, specifications, etc.*

Typical Issues

Was the problem caused by a failure to communicate? Did a method or system exist for communicating between the groups or individuals? Was an error caused by misunderstood communication between personnel? Was there incorrect, incomplete, or otherwise inadequate communication between workers during a shift or between workers during a shift change? Was there a problem communicating with contractors or customers?

Typical Recommendations

- Provide a backup means of communication when the primary system is inoperable
- Establish standard terminology for equipment and operations
- Use the repeat-back method of communication
- Conduct shift change meetings to alert oncoming shifts of special job tasks, safety issues, or problems that occurred during the previous shift

Examples

An operator opened the wrong valve, resulting in a process upset. He misunderstood the verbal instructions from a coworker. No repeat-back or other verification method was used.

A tank transfer was in progress during shift change. During shift change, the shift going off duty did not tell the one coming on duty that the transfer was in progress. The tank overflowed.

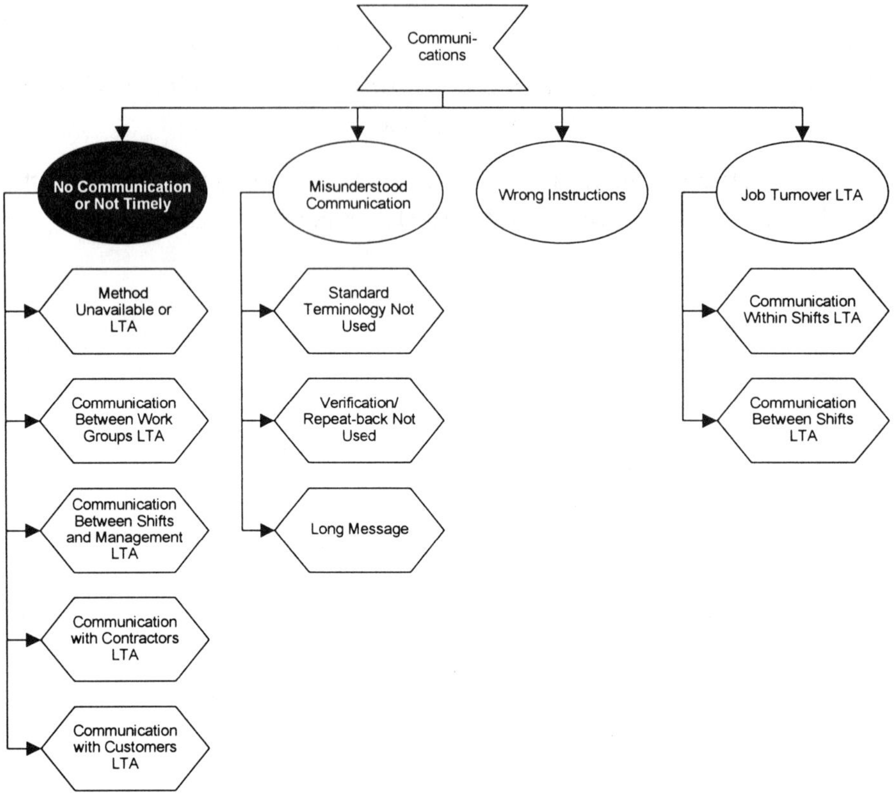

Typical Issues

Was the problem caused by a failure to communicate? Did a method or system exist for communicating between the groups or individuals? Did the communication take place too late? Did obstacles hinder or delay communication?

Note: *Each individual involved in the occurrence should be questioned regarding messages he or she feels should have been received or transmitted. Determine what means of communication were used (i.e., the techniques). Persons on all sides of a communication link should be questioned regarding known or suspected problems.*

Typical Recommendations

- Provide a backup means of communication when the primary system is inoperable
- Establish formal means of communication when required
- Conduct meetings between shift workers and management

Example

An operator failed to close a valve when needed, resulting in a process upset. He should have received an instruction from control room personnel to close the valve. The instruction was not given to the operator in time because the two-way radios did not work in the area in which the operator was located.

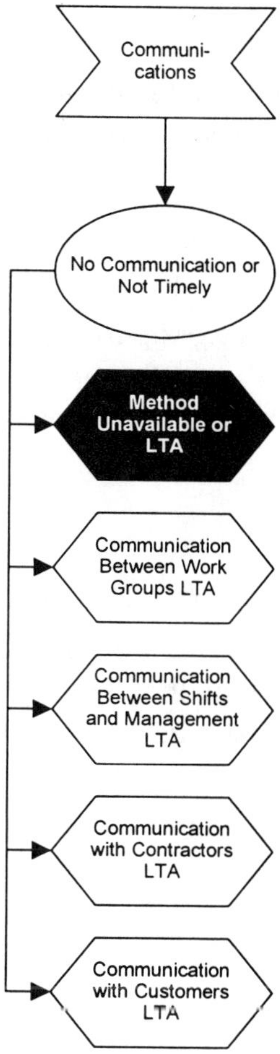

Typical Issues

Did a method or system exist for communicating the necessary message or information? Was the communication system out of service or otherwise unavailable at the time of the incident?

Typical Recommendations

- Ensure that some method of communication is in working order at all times
- When the primary method of communication is unavailable, provide some temporary means of communication (e.g., two-way radios)

Example

An automatic valve was stuck open. The control room operator attempted to contact the building operator using the public address system to instruct him to manually close the valve. The public address system was not functioning properly, and the building operator could not be contacted, resulting in overflow of a vessel.

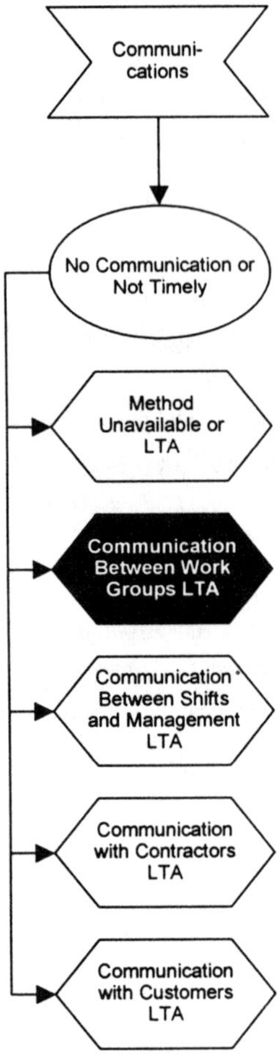

Typical Issues

Did lack of communication between work groups (production, technical, maintenance, warehouse, etc.) contribute to the incident? Did methods exist for communicating between work groups?

Typical Recommendation

- Establish an administrative procedure that requires a work permit (or sign-in/sign-out) to be issued and authorized by the controlling work group before another work group may perform job tasks in the controlling work group's area

Example

A tank overflowed because maintenance had taken the liquid level instrumentation out of service for calibration. A misunderstanding with production occurred over which equipment was out of service. Believing that it was another instrument that was being calibrated, production started a transfer into the tank, resulting in an overflow.

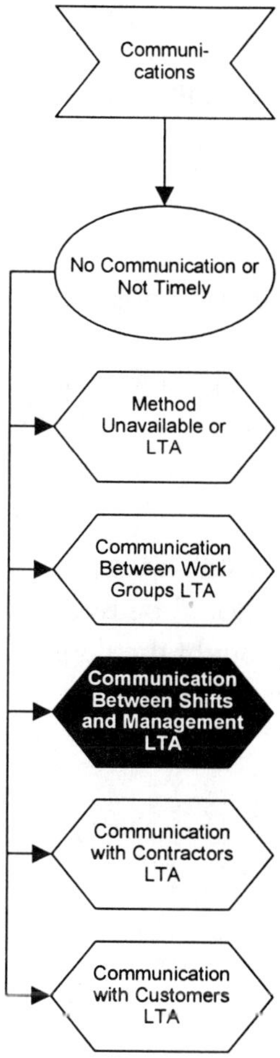

Typical Issues

Did lack of communication between management and the shifts contribute to the incident? Had management effectively communicated policies to the employees? Were employees' concerns communicated to management?

Typical Recommendations

- Conduct shift meetings that involve members of management during various shifts
- Provide a management/shift logbook for each production unit so that various issues concerning production, safety, maintenance, etc., can be communicated as needed between management and all shifts (and between shifts)
- Provide a suggestion box in the facility
- Encourage all employees to submit work requests/suggestions regarding maintenance, safety issues, etc., to management. Post the work requests weekly to acknowledge that the request was received, and include the status of the request

Examples

A valve failed, resulting in a process upset. Shift employees had noticed problems with the valve and had expressed concern to the first-line supervisors, but the problem had not been recognized by management and corrected.

A policy was recently changed that required personnel to enter their work hours into a new computer system. However, the details of the policy were not communicated to personnel on the weekend shift.

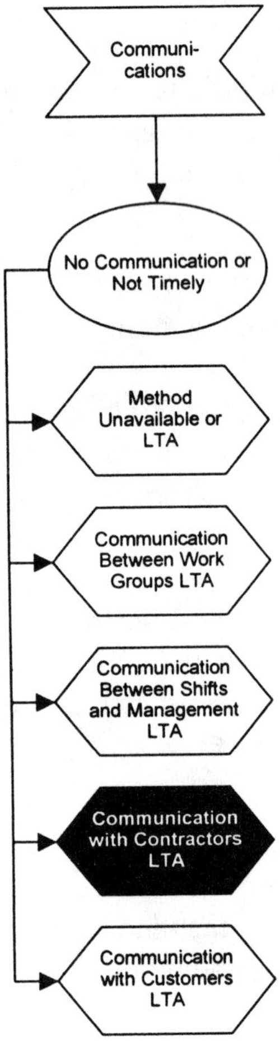

Typical Issues

Were there problems communicating with contractors? Were they made aware of changes in policies and procedures?

Typical Recommendations

- Provide methods for communication between your company and contractors
- Ensure that contractors have a designated contact in your organization
- Provide policy and procedure updates to contractors

Example

A contractor was assigned the task of digging a ditch to install an underground tank. An engineer used flags to mark areas that should be avoided so as not to disturb underground utilities. The contractor thought the flags marked the spot to dig. As a result, a natural gas line feeding the plant was struck and broken.

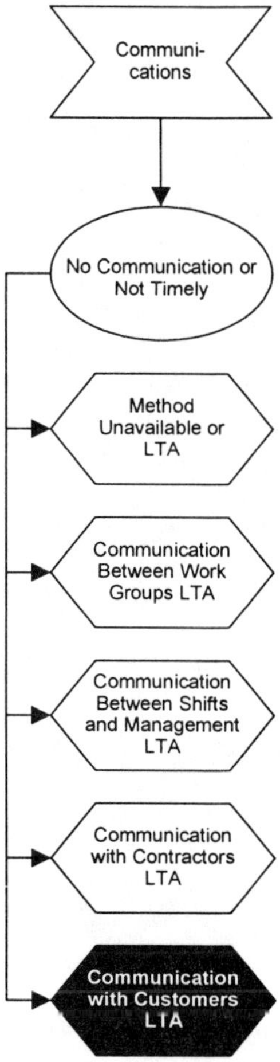

Typical Issues

Were there problems communicating with customers? Are customers able to communicate their needs to the company? Does the company respond to requests from customers? Are phone calls returned and letters acknowledged?

Typical Recommendations

- Track customer calls and letters to ensure timely processing
- Provide convenient means for customers to contact the company

Examples

A customer called with a question on installation of your product. The customer was told that an installation technician would return the call in an hour. The customer called again the next day after having not been called back.

A customer called in to order a batch of stainless steel bolts. The order was not recorded correctly, and the customer received a batch of threaded rods instead.

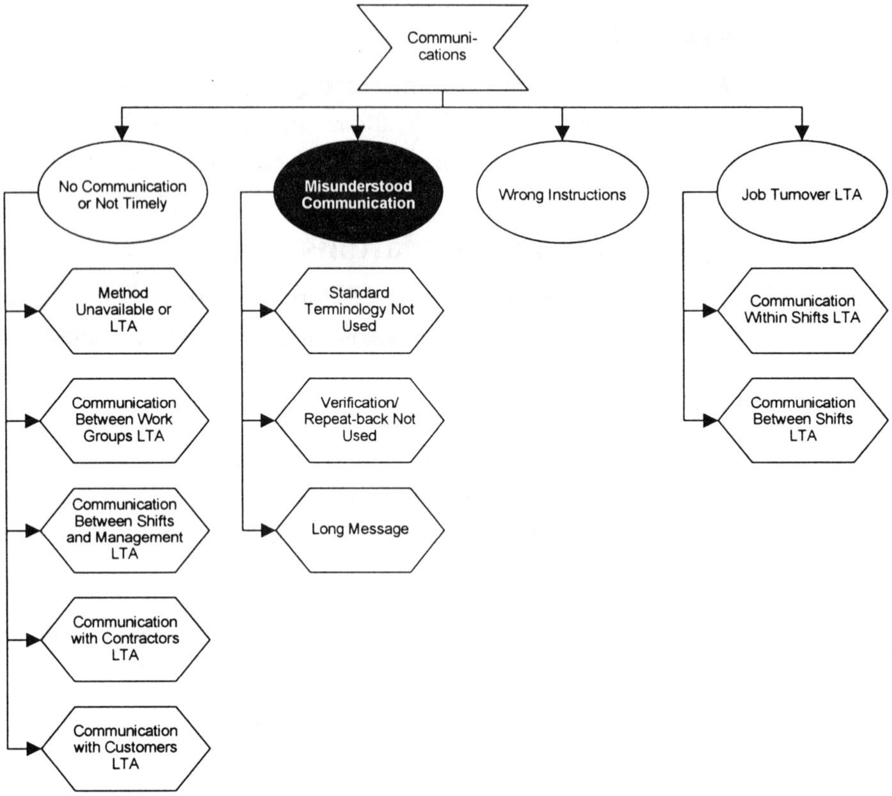

Typical Issues

Was an error caused by misunderstood communications between personnel? Was there an error in verbal communication? Did someone misunderstand a hand signal? Was a sign misunderstood? Were oral instructions given when written instructions should have been provided?

Typical Recommendations

- Establish standard terminology for equipment and operations
- Use the repeat-back method of communication
- Provide written instructions when necessary
- Minimize interference from noise

Example

An operator located in a noisy part of the plant was given an instruction by "walkie-talkie" to open Valve B-2. He thought the verbal instruction was to open Valve D-2. No repeat-back or other type of verification was used. He opened D-2, resulting in a process upset.

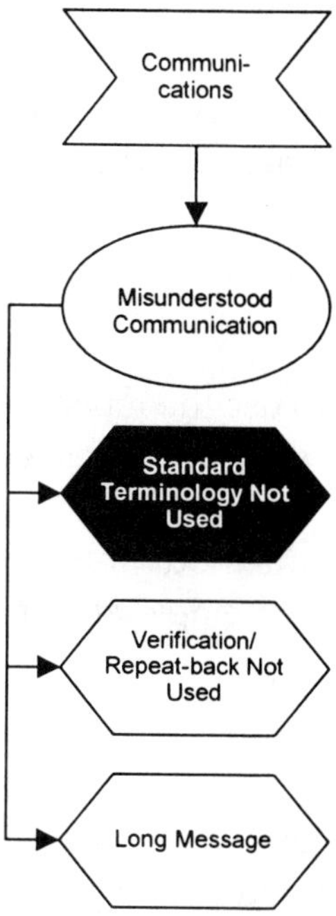

Typical Issues

Was standard or accepted terminology used? Could the communication be interpreted more than one way? Did one piece of equipment have two or more commonly used names? Could the terminology have applied to more than one item?

Typical Recommendations

- Establish standard terminology for equipment, process operations, and maintenance operations. Encourage all employees to stop using non-standard terminology
- Avoid ambiguous terms and phrases in procedures, work instructions, logbooks, etc.

Example

An operator was told to verify that a solution was clear before adding it to a process. The operator thought that "clear" meant "not cloudy." What was actually meant was "no color" since color was an indication of contaminants in the solution. The solution was clear (translucent), but had a slightly pink tint. As a result, an out-of-specification solution was used.

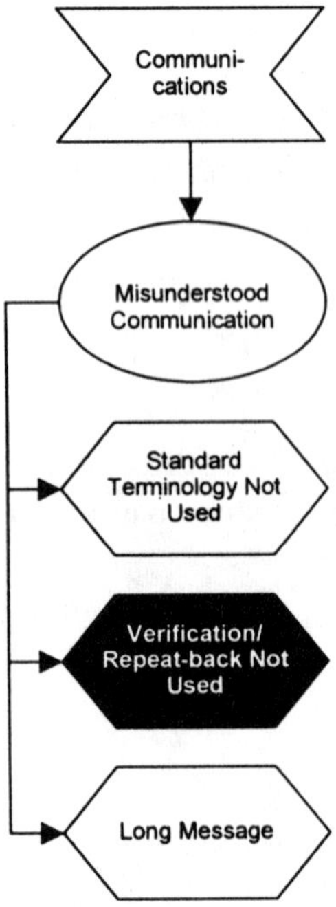

Typical Issue

Was a communication error caused by failure to repeat a message back to the sender for the purpose of verifying that the message was heard and understood correctly?

Typical Recommendations

- Encourage employees and personnel at all levels to use the repeat-back communication method to ensure thorough understanding of related job tasks
- If employees/workers forget to use the repeat-back method, instruct supervisors and work team leaders to request that the employee repeat back

Example

An operator was given an instruction by "walkie-talkie" to open a valve. The instruction was to open Valve B-2. The operator understood D-2. No repeat-back or other type of verification was used.

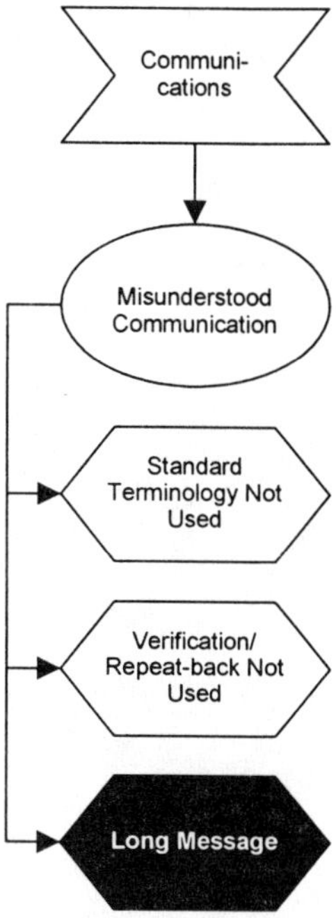

Typical Issues

Was a message or instruction misunderstood because it was too long? Should the message have been written instead of spoken? Could the message have been shortened or broken up?

Typical Recommendations

- Keep oral instructions short and rehearsed (especially if communicating in noisy areas)
- If several lengthy details must be conveyed, consider providing them as written instructions rather than oral (i.e., generate a written procedure)

Example

An operator was verbally instructed to open Valves A-7, B-4, B-5, C-6, D-6, D-7, D-8, and F-1. He failed to open D-6, resulting in a process upset. No written instructions were given.

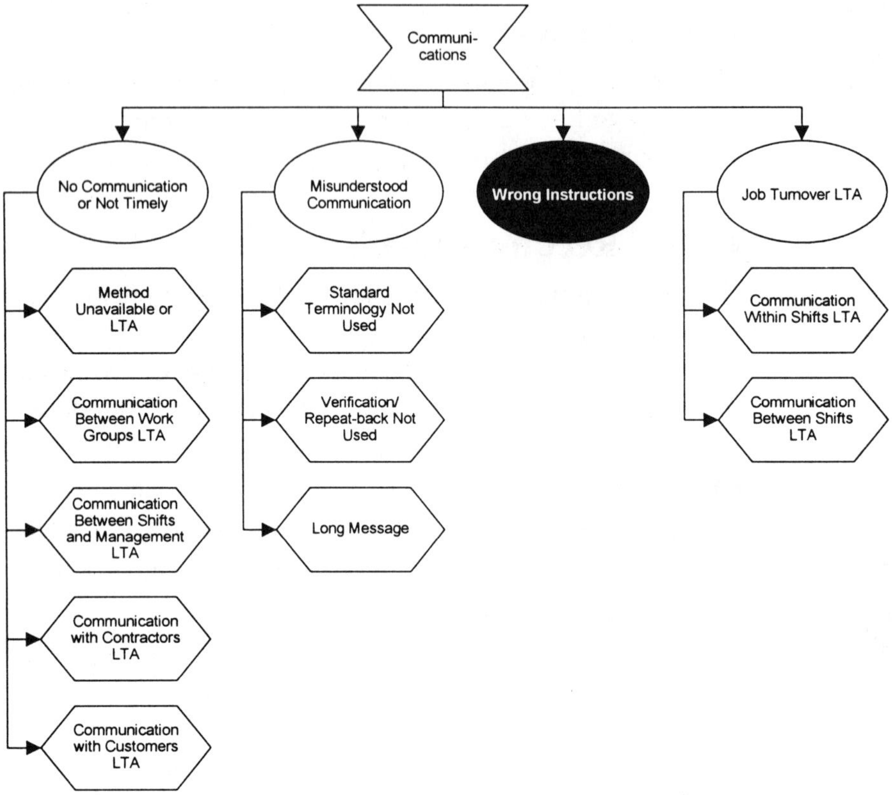

Typical Issue

Was the communication errant or inaccurate?

Typical Recommendation

- Consult the procedure, training, supervision, human factors engineering, and/or personal performance branches of the map

Example

A supervisor told an operator to open Valve 101 instead of Valve 201.

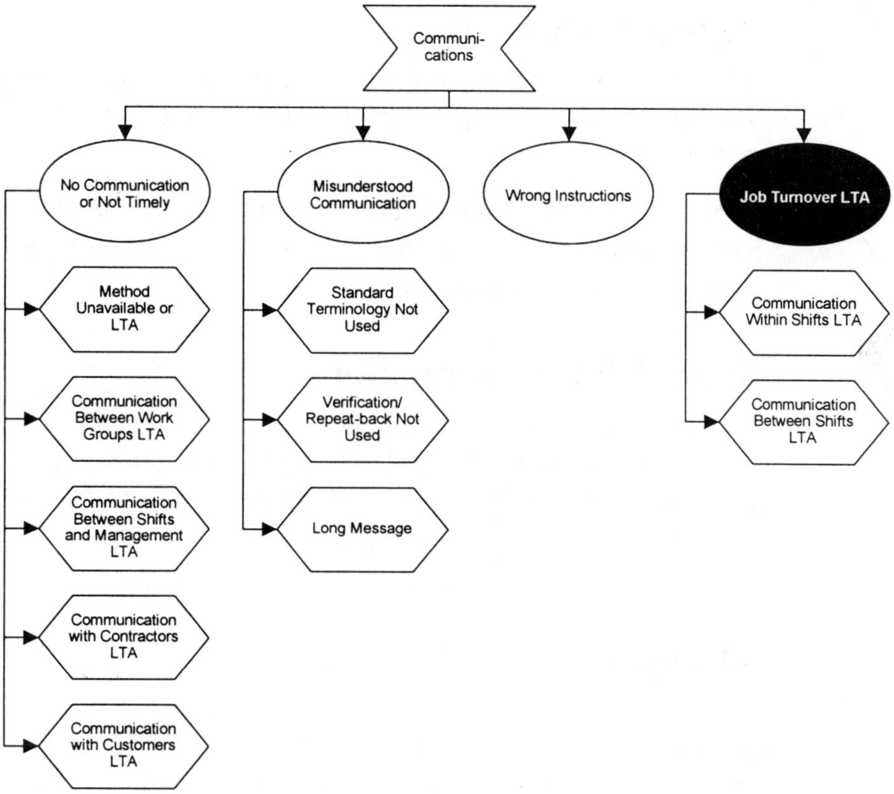

Typical Issue

Was there incorrect, incomplete, or otherwise inadequate communication between workers during a shift or between shifts during a shift change?

Typical Recommendations

- Provide a means of communication for employees working on the same shift (e.g., two-way radios)
- Conduct shift change meetings to alert oncoming shifts of special job tasks, safety issues, or problems that occurred during the previous shift
- Provide guidance on the content of shift turnovers

Example

A tank transfer was in progress during shift change. During shift change, the shift going off duty did not tell the one coming on that the transfer was in progress. The tank overflowed.

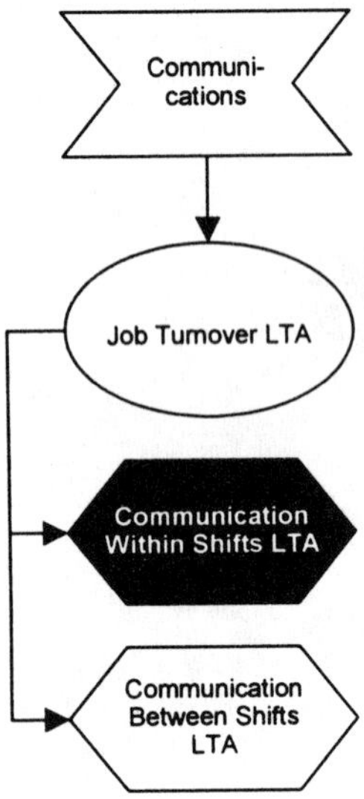

Typical Issues

Was there incorrect, incomplete, or otherwise inadequate communication between workers during a shift? Could a more effective method of communication have been used?

Note: *Planning and coordination of jobs between individuals and work groups should be coded under* Teamwork LTA (Immediate Supervision, Supervision During Work).

Typical Recommendations

- Encourage employees to alert others on their shift of changes in job tasks that may affect others (tell others when you plan to take a break, tell others when you move from one job location to another, etc.)
- Encourage employees to keep each other informed about changes in equipment status that may affect other areas of the plant

Example

A tank transfer was in progress when Operator A went on break. He mentioned to Operator B that the transfer was going on, but Operator B did not realize that he needed to stop the transfer. As a result, the tank overflowed.

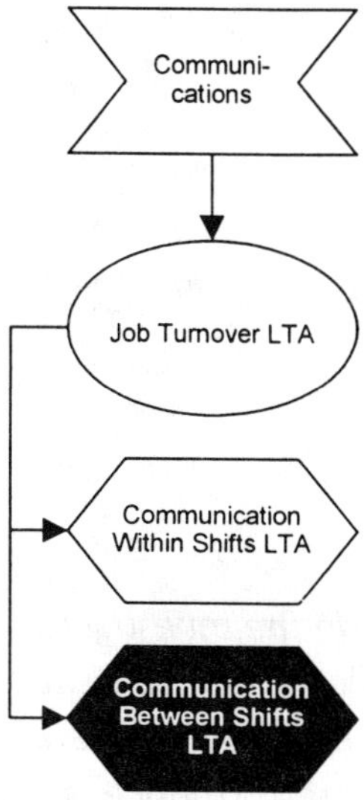

Typical Issue

Was there incorrect, incomplete, or otherwise inadequate communication between workers during a shift change?

Typical Recommendations

- Conduct shift change meetings to alert the oncoming shifts of special job tasks, safety issues, or problems that occurred during the previous shift
- Use logbooks to communicate between shifts
- Provide guidance on the content of shift turnovers

Example

A tank transfer was in progress during shift change. During the turnover, the shift going off duty did not tell the one coming on that the transfer was in progress. The tank overflowed.

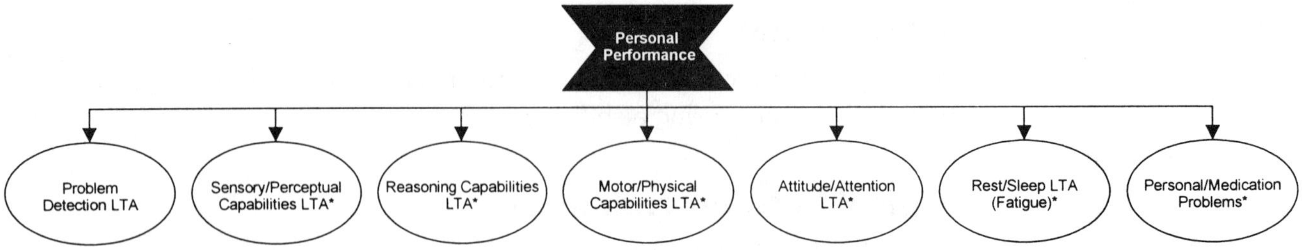

* Do not include these nodes in the investigation report.

Typical Issues

Note: *The six asterisked Level D nodes (ovals, near root causes) are included to provide the investigator with an understanding of the types of problems that might be categorized as* Personal Performance *issues. However, the investigator should not include these Level D nodes in the investigation report. Also, there should be management systems in place to detect and correct most (if not all) personal performance issues* before *a loss event occurs. Therefore, the failure or absence of the management systems should be coded as well.*

Did the worker's physical or mental well-being, attitude, mental capacity, attention span, rest, substance abuse, etc., adversely affect the performance of the task? Was the problem the result of the individual not being capable of performing the task or not wanting to do his or her job? Was a personal performance problem promptly detected? Was corrective action promptly taken?

Typical Recommendations

- Ensure that there is a process in place to detect personal performance problems
- Provide a means for personnel to self-report problems

Examples

An operator failed to close a valve after completion of a transfer. The operator was not paying attention to the level of the tank into which the material was being transferred. The operator had a history of not paying attention to his work. He had been involved in several other incidents during which he had left his job site or was not performing his job requirements. Other operators performed these same job requirements with no problems.

An individual came to work drunk. The operator was stumbling while walking to his workstation. However, no one did anything to stop him from going to work.

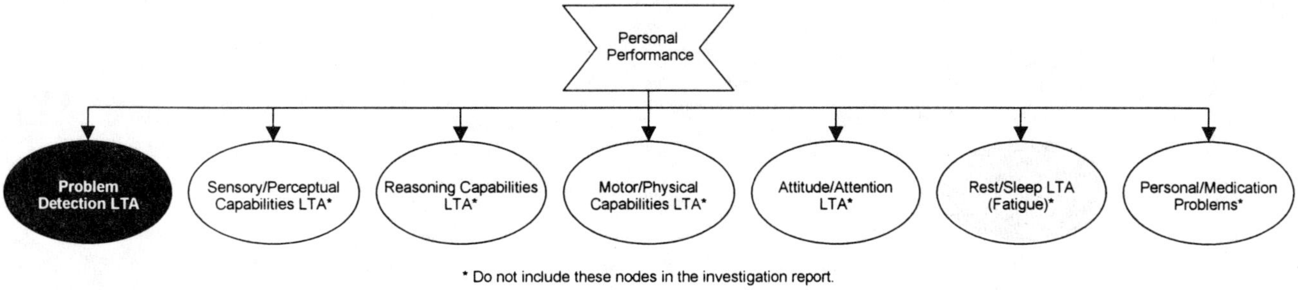

Typical Issues

Did personal performance issues contribute to the event? Should the personal performance issues have been detected prior to the event?

Note: *Consider dual coding under* Supervision LTA (Immediate Supervision, Supervision During Work).

Typical Recommendations

- Provide supervisors with training on the detection of personal problems
- Give supervisors the authority to remove workers from hazardous assignments when personal problems are detected
- Encourage coworkers to help identify personal performance problems

Examples

A worker came to work drunk. He was having trouble walking and talking. While going to get a part from the warehouse, he fell down some steps and injured himself and another worker.

Six months ago, a maintenance technician was hired who could not read. His supervisor had not detected the problem, even though this technician had trouble with all of his nonroutine tasks (those that required him to use a procedure).

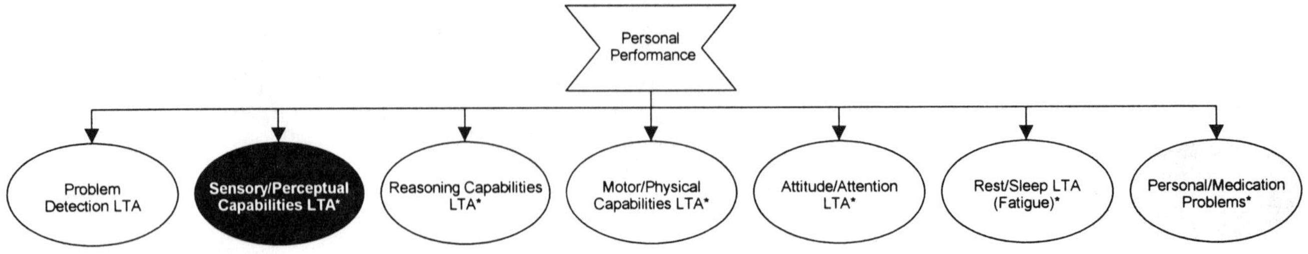

* Do not include these nodes in the investigation report.

Typical Issues

Note: *Code as* Personal Performance *only. The six asterisked Level D nodes (ovals, near root causes) are included to provide the investigator with an understanding of the types of problems that might be categorized as* Personal Performance *issues. However, the investigator should not include these Level D nodes in the investigation report. Also, there should be management systems in place to detect and correct most (if not all) personal performance issues* before *a loss event occurs. Therefore, the failure or absence of the management systems should be coded as well.*

Was the problem a result of less than adequate vision (e.g., poor visual acuity, color blindness, tunnel vision)? Was the problem a result of some defect in hearing (e.g., hearing loss, tone deafness)? Was the problem a result of some sensory defect (e.g., poor sense of touch or smell)?

Typical Recommendations

- Ensure that job requirements are complete, including required physical/perceptual capabilities
- Provide reasonable accomodations for coworkers with sensory/perceptual limits

Note: *A review of the human factors engineering for the process is also appropriate to accommodate a wider spectrum of sensory capabilities. For example: Can the displays be redesigned so that lights that indicate "closed" conditions of valves are always in the same relative location on the panel? Can more chart recorders be installed with fewer points per chart?*

Example

An operator read the wrong temperature on a chart that recorded temperatures for several tanks. The chart was color coded. The operator was partially color blind and confused the readings. He recorded a temperature that was in range when the actual temperature was out of range.

Note: *Consider coding under* Employee Screening/Hiring LTA (Administrative/Management Systems, SPACs LTA) *because there should be management controls to ensure employees possess the required job capabilities.*

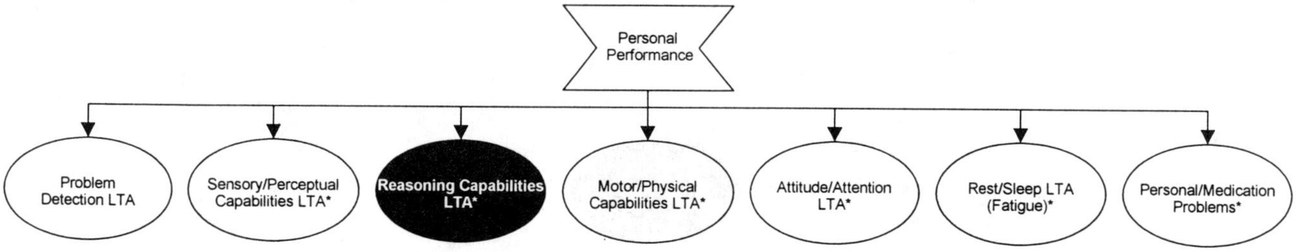

* Do not include these nodes in the investigation report.

Typical Issues

Note: *Code as* Personal Performance *only. The six asterisked Level D nodes (ovals, near root causes) are included to provide the investigator with an understanding of the types of problems that might be categorized as* Personal Performance *issues. However, the investigator should not include these Level D nodes in the investigation report. Also, there should be management systems in place to detect and correct most (if not all) personal performance issues* <u>*before*</u> *a loss event occurs. Therefore, the failure or absence of the management systems should be coded as well.*

Was the problem caused by inadequate intellectual capacity? Does the person frequently make wrong decisions? In general, does the person have difficulty processing information? Do other workers have difficulty performing these tasks or is it isolated to this one worker?

Typical Recommendation

- Review employee screening and hiring processes to ensure that the individuals who are hired have the required reasoning capabilities

Examples

An operator made a mistake in a calculation and added too much material to the mixer. The operator had frequently made errors with calculations and appeared to have problems with numbers. Other operators did not have difficulty performing these tasks.

An operator missed several steps in a procedure. The operator was unable to clearly understand the procedures because they were written at a sixth-grade level and he could only read at a second-grade level.

Note: *Consider coding under* Employee Screening/Hiring LTA (Administrative/Management Systems, SPACs LTA) *as well, since there should be management controls to ensure that employees possess the appropriate reading and mathematical skills.*

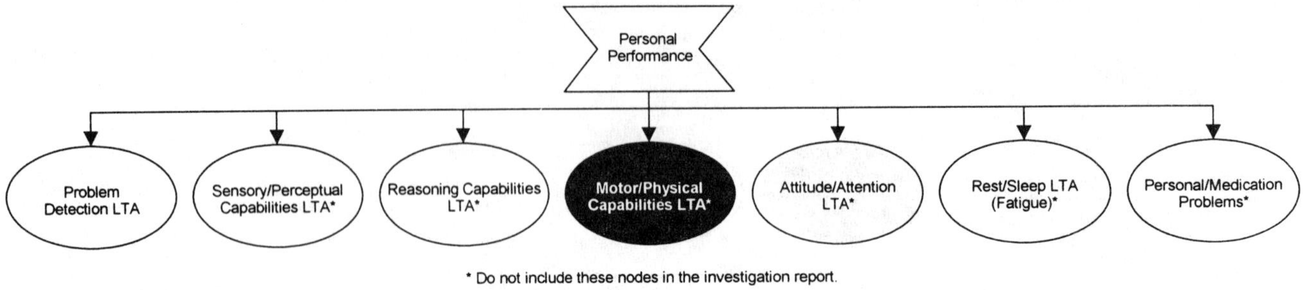

Typical Issues

Note: *Code as* Personal Performance *only. The six asterisked Level D nodes (ovals, near root causes) are included to provide the investigator with an understanding of the types of problems that might be categorized as* Personal Performance *issues. However, the investigator should not include these Level D nodes in the investigation report. Also, there should be management systems in place to detect and correct most (if not all) personal performance issues* <u>*before*</u> *a loss event occurs. Therefore, the failure or absence of the management systems should be coded as well.*

Can the causal factor be attributed to trouble with inadequate coordination or inadequate strength? Was the problem a result of inadequate size or stature of the individual involved? Did other physical limitations (e.g., shaking, poor reaction time) contribute to the problem?

Typical Recommendations

- Ensure that job requirements are complete, including required physical/perceptual capabilities
- Provide reasonable accomodations for workers with physical limitations

Note: *A review of the human factors engineering for the process is also appropriate. Is it reasonable for an "average" individual to perform this task? Can the individual be provided with a tool to assist in the task? Can the task be redesigned to reduce the physical requirements?*

Example

A tank overflowed because the operator could not close the valve. The valve was large and difficult to close. The operator did not have the strength to close the valve. By the time he obtained help in closing it, the tank had overflowed.

Note: *Consider coding under* Employee Screening/Hiring LTA (Administrative/Management Systems, SPACs LTA) *because there should be management controls to ensure employees possess the required job capabilities.*

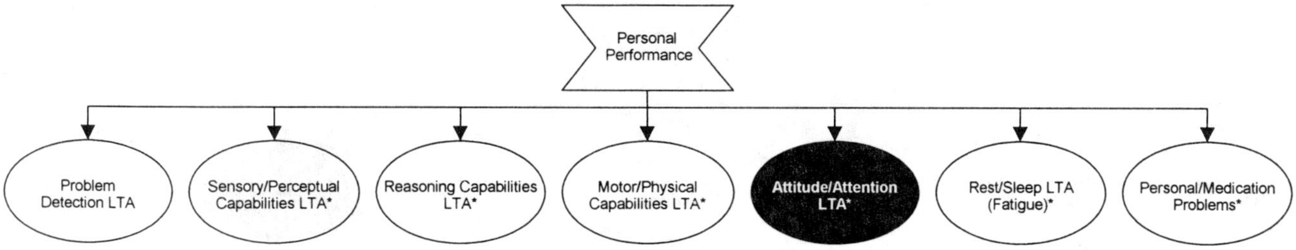

* Do not include these nodes in the investigation report.

Typical Issues

Note: *Code as* Personal Performance *only. The six asterisked Level D nodes (ovals, near root causes) are included to provide the investigator with an understanding of the types of problems that might be categorized as* Personal Performance *issues. However, the investigator should not include these Level D nodes in the investigation report. Also, there should be management systems in place to detect and correct most (if not all) personal performance issues before a loss event occurs. Therefore, the failure or absence of the management systems should be coded as well.*

Was the problem a result of a poor attitude on the part of an individual? Did the individual involved show signs of emotional illness? Was the problem caused by lack of attention? Does the individual involved in this occurrence frequently "daydream"? Is the person distracted easily? Is the person's ability to maintain vigilance frequently below minimum acceptable standards? Do other workers have difficulty performing these tasks or is it isolated to this one worker?

Typical characteristics include:

- engages in horseplay
- is not at work location
- does not perform expected work
- exhibits maliciousness
- exhibits inability to operate under stress
- exhibits insubordination
- exhibits inability to work well or communicate with other people
- ignores safety rules

Examples

An operator failed to close a valve while filling a tank, resulting in an overflow from the tank and a process upset. The operator often was away from his assigned work location for personal reasons, such as making personal phone calls.

An operator failed to stop a transfer, resulting in a tank overflow. The operator had a history of being distracted easily and losing track of the next step in the process.

Note: *Consider coding under* Employee Screening/Hiring LTA (Administrative/Management Systems, SPACs LTA) *because there should be management controls to ensure employees possess the required job capabilities. Also consider coding under* Supervision LTA (Immediate Supervision, Supervision During Work) *because supervision should detect this problem.*

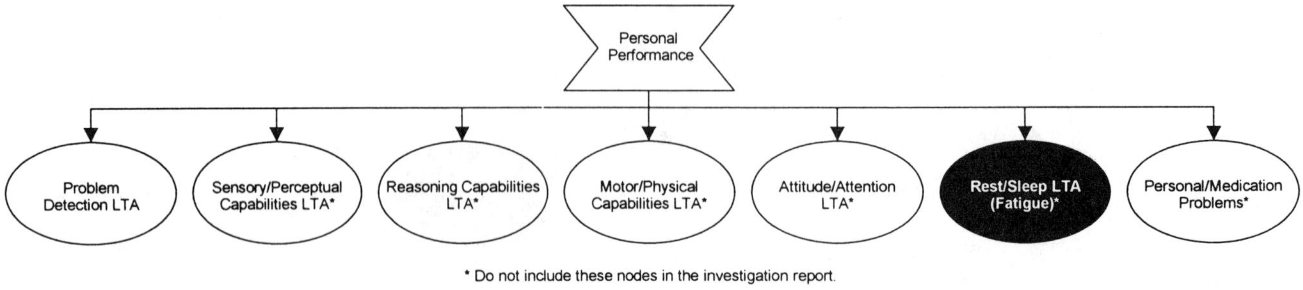

Typical Issues

Note: *Code as* Personal Performance *only. The six asterisked Level D nodes (ovals, near root causes) are included to provide the investigator with an understanding of the types of problems that might be categorized as* Personal Performance *issues. However, the investigator should not include these Level D nodes in the investigation report. Also, there should be management systems in place to detect and correct most (if not all) personal performance issues* <u>*before*</u> *a loss event occurs. Therefore, the failure or absence of the management systems should be coded as well.*

Note: *This node addresses problems associated with an individual's rest and sleep practices outside of the workplace. Problems with workers who are forced to work unreasonable amounts of overtime should be coded using the* Immediate Supervision *or* Administrative/Management Systems *segments of the map.*

Was the worker involved in the incident asleep while on duty? Was the person too tired to perform the job?

Example

A mechanic was found asleep while he was supposed to be calibrating equipment. The mechanic had another job away from the site and routinely appeared to be extremely tired.

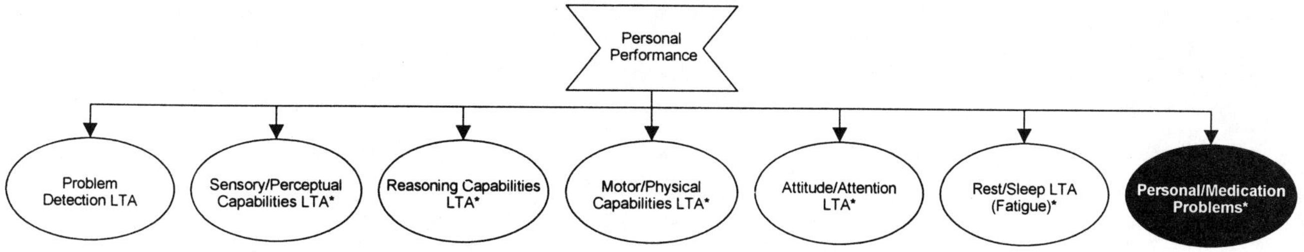

* Do not include these nodes in the investigation report.

Typical Issues

Note: *Code as* Personal Performance *only. The six asterisked Level D nodes (ovals, near root causes) are included to provide the investigator with an understanding of the types of problems that might be categorized as* Personal Performance *issues. However, the investigator should not include these Level D nodes in the investigation report. Also, there should be management systems in place to detect and correct most (if not all) personal performance issues* ***before*** *a loss event occurs. Therefore, the failure or absence of the management systems should be coded as well.*

Is the individual experiencing personal problems that are affecting his or her job performance? Is the individual taking medications that affect his or her job performance?

Typical symptoms include:

- chronic inattention
- acute inattention
- frequent daydreaming
- easily distracted
- poor vigilance
- illness
- impairment due to prescription drugs
- poor psychological health
- abuse of drugs/alcohol

Typical Recommendations

- Establish an employee assistance program
- Inform and encourage workers to take advantage of employee assistance programs

Example

An operator was prescribed a medication that caused drowsiness. During a tank transfer, he lost track of time and the tank overflowed.

ROOT CAUSE MAP™

11 x 17 inch version of the Root Cause Map with cross references to the node descriptions in Appendix A of this handbook.

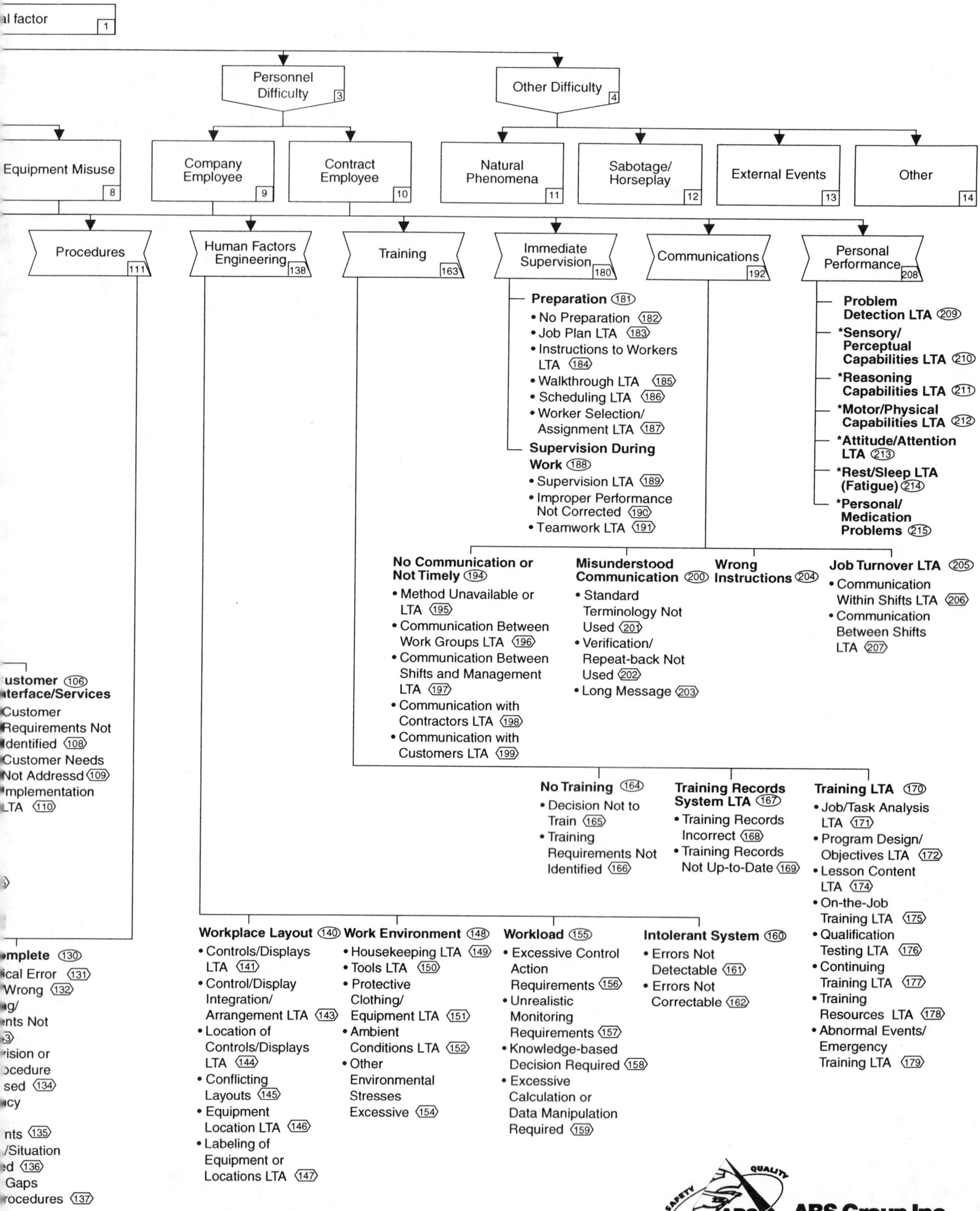

ABS Group Inc.
Risk & Reliability Division

*Note: These nodes are for descriptive purposes only. Code only to Personal Performance (Node 208).

Root Cause Map™

Start here with each caus

Equipment Difficulty 2

- Equipment Design Problem 5
  - Design Input/Output 15
    - **Design Input LTA** 16
    - **Design Output LTA** 17
- Equipment Reliability Program Problem 6
  - Equipment Records 18
    - **Equipment Design Records LTA** 19
    - **Equipment Operating/ Maintenance History LTA** 20
  - Equipment Reliability Program Design LTA 21
    - **No Program** 22
    - **Program LTA** 23
      - Analysis/Design Procedure LTA 24
      - Inappropriate Type of Maintenance Assigned 25
      - Risk Acceptance Criteria LTA 26
      - Allocation of Resources LTA 27
  - Equipment Reliability Program Implementation LTA 28
    - **Corrective Maintenance LTA** 29
      - Troubleshooting/ Corrective Action LTA 30
      - Repair Implementation LTA 31
    - **Preventive Maintenance LTA** 32
      - Frequency LTA 33
      - Scope LTA 34
      - Activity Implementation LTA 35
    - **Predictive Maintenance LTA** 36
      - Detection LTA 37
      - Monitoring LTA 38
      - Troubleshooting/ Corrective Action LTA 39
      - Activity Implementation LTA 40
    - **Proactive Maintenance LTA** 41
      - Event Specification LTA 42
      - Monitoring LTA 43
      - Scope LTA 44
      - Activity Implementation LTA 45
    - **Failure Finding Maintenance LTA** 46
      - Frequency LTA 47
      - Scope LTA 48
      - Troubleshooting/ Corrective Action LTA 49
      - Repair Implementation 50
    - **Routine Equipment Rounds LTA** 51
      - Frequency LTA 52
      - Scope LTA 53
      - Activity Implementation LTA 54
- Installation/ Fabrication 7
  - Administrative/ Management Systems 55

Note: Node numbers correspond to matching page in Appendix A of the *Root Cause Analysis Handbook*

Standards, Policies, or Administrative Controls (SPACs) LTA 57
- No SPACs 59
- Not Strict Enough 60
- Confusing, Contradictory, or Incomplete 61
- Technical Error 62
- Responsibility for Item/Activity Not Adequately Defined 63
- Planning, Scheduling, or Tracking of Work Activities LTA 64
- Rewards/ Incentives LTA 65
- Employee Screening/Hiring LTA 66

Standards, Policies, or Administrative Controls (SPACs) Not Used 67
- Communication of SPACs LTA 69
- Recently Changed 70
- Enforcement LTA 71

Safety/Hazard/Risk Review 72
- Review LTA or Not Performed 74
- Recommendations Not Yet Implemented 75
- Risk Acceptance Criteria LTA 76
- Review Procedure LTA 77

Problem Identification/ Control 78
- Problem Reporting LTA 80
- Problem Analysis LTA 81
- Audits LTA 82
- Corrective Action LTA 83
- Corrective Actions Not Yet Implemented 84

Product/Material Control 85
- Handling LTA 87
- Storage LTA 88
- Packaging/ Shipping LTA 89
- Unauthorized Material Substitution 90
- Product Acceptance Criteria LTA 91
- Product Inspections LTA 92

Procurement Control 93
- Purchasing Specifications LTA 95
- Control of Changes to Procurement Specifications LTA 96
- Material Acceptance Requirements LTA 97
- Material Inspections LTA 98
- Contractor Selection LTA 99

Document and Configuration Control 100
- Change Not Identified 102
- Verification of Design/Field Changes LTA (No PSSR) 103
- Documentation Content Not Kept Up-to-Date 104
- Control of Official Documents LTA 10

Shape	Description
(pentagon)	Primary Difficulty Source
(rectangle)	Problem Category
(notched rectangle)	Root Cause Category
(ellipse)	Near Root Cause
(hexagon)	Root Cause

Note: LTA = Less Than Adequate

Not Used 112
- Not Available or Inconvenient to Obtain 113
- Procedure Difficult to Use 114
- Use Not Required but Should Be 115
- No Procedure for Task 116

Misleading/Confusing 117
- Format Confusing or LTA 118
- More Than One Action per Step 120
- No Checkoff Space Provided but Should Be 121
- Inadequate Checklist 122
- Graphics LTA 123
- Ambiguous or Confusing Instructions/Requirements 124
- Data/Computations Wrong/ Incomplete 125
- Insufficient or Excessive References 126
- Identification of Revised Steps LTA 127
- Level of Detail LTA 128
- Difficult to Identify 129

Wrong/Inc
- Typograph
- Sequence
- Facts Wro Requireme Correct 1
- Wrong Re Expired P Revision L
- Inconsiste Between Requireme
- Incomplete Not Cover
- Overlap or Between F

W:\Graphics Files\Manuals\Root Cause Analysis Handbook\Figures other than the root cause map\Page number root cause map.vsd 08/05/99 15:16

Government Institutes Mini-Catalog

PC #	ENVIRONMENTAL TITLES	Pub Date	Price
629	ABCs of Environmental Regulation: Understanding the Fed Regs	1998	$49
627	ABCs of Environmental Science	1998	$39
585	Book of Lists for Regulated Hazardous Substances, 8th Edition	1997	$79
579	Brownfields Redevelopment	1998	$79
4088	CFR Chemical Lists on CD ROM, 1997 Edition	1997	$125
4089	Chemical Data for Workplace Sampling & Analysis, Single User Disk	1997	$125
512	Clean Water Handbook, 2nd Edition	1996	$89
581	EH&S Auditing Made Easy	1997	$79
587	E H & S CFR Training Requirements, 3rd Edition	1997	$89
4082	EMMI-Envl Monitoring Methods Index for Windows-Network	1997	$537
4082	EMMI-Envl Monitoring Methods Index for Windows-Single User	1997	$179
525	Environmental Audits, 7th Edition	1996	$79
548	Environmental Engineering and Science: An Introduction	1997	$79
643	Environmental Guide to the Internet, 4rd Edition	1998	$59
560	Environmental Law Handbook, 14th Edition	1997	$79
353	Environmental Regulatory Glossary, 6th Edition	1993	$79
625	Environmental Statutes, 1998 Edition	1998	$69
4098	Environmental Statutes Book/CD-ROM, 1998 Edition	1997	$208
4994	Environmental Statutes on Disk for Windows-Network	1997	$405
4994	Environmental Statutes on Disk for Windows-Single User	1997	$139
570	Environmentalism at the Crossroads	1995	$39
536	ESAs Made Easy	1996	$59
515	Industrial Environmental Management: A Practical Approach	1996	$79
510	ISO 14000: Understanding Environmental Standards	1996	$69
551	ISO 14001: An Executive Report	1996	$55
588	International Environmental Auditing	1998	$149
518	Lead Regulation Handbook	1996	$79
478	Principles of EH&S Management	1995	$69
554	Property Rights: Understanding Government Takings	1997	$79
582	Recycling & Waste Mgmt Guide to the Internet	1997	$49
603	Superfund Manual, 6th Edition	1997	$115
566	TSCA Handbook, 3rd Edition	1997	$95
534	Wetland Mitigation: Mitigation Banking and Other Strategies	1997	$75

PC #	SAFETY and HEALTH TITLES	Pub Date	Price
547	Construction Safety Handbook	1996	$79
553	Cumulative Trauma Disorders	1997	$59
559	Forklift Safety	1997	$65
539	Fundamentals of Occupational Safety & Health	1996	$49
612	HAZWOPER Incident Command	1998	$59
535	Making Sense of OSHA Compliance	1997	$59
589	Managing Fatigue in Transportation, *ATA Conference*	1997	$75
558	PPE Made Easy	1998	$79
598	Project Mgmt for E H & S Professionals	1997	$59
552	Safety & Health in Agriculture, Forestry and Fisheries	1997	$125
613	Safety & Health on the Internet, 2nd Edition	1998	$49
597	Safety Is A People Business	1997	$49
463	Safety Made Easy	1995	$49
590	Your Company Safety and Health Manual	1997	$79

GOVERNMENT INSTITUTES ORDER FORM

4 Research Place, Suite 200 • Rockville, MD 20850-3226
Tel (301) 921-2323 • Fax (301) 921-0264
Internet: http://www.govinst.com • E-mail: giinfo@govinst.com

3 EASY WAYS TO ORDER

1. **Phone:** **(301) 921-2323**
   Have your credit card ready when you call.
2. **Fax:** **(301) 921-0264**
   Fax this completed order form with your company purchase order or credit card information.
3. **Mail:** **Government Institutes**
   4 Research Place, Suite 200
   Rockville, MD 20850-3226 USA
   Mail this completed order form with a check, company purchase order, or credit card information.

PAYMENT OPTIONS

❑ **Check** *(payable to Government Institutes in US dollars)*

❑ **Purchase Order** *(This order form must be attached to your company P.O. Note: All International orders must be prepaid.)*

❑ **Credit Card** ☐ VISA ☐ MasterCard ☐ American Express

Exp.___/___

Credit Card No. ______________________________

Signature ______________________________

(Government Institutes' Federal I.D.# is 13-2695912)

CUSTOMER INFORMATION

Ship To: (Please attach your purchase order)

Name: ______________________________

GI Account # *(7 digits on mailing label)*: ______________________________

Company/Institution: ______________________________

Address: ______________________________
(Please supply street address for UPS shipping)
____

City: ______________ State/Province: ______________

Zip/Postal Code: ______________ Country: ______________

Tel: (   ) ______________________________

Fax: (   ) ______________________________

Email Address: ______________________________

Bill To: (if different from ship-to address)

Name: ______________________________

Title/Position: ______________________________

Company/Institution: ______________________________

Address: ______________________________
(Please supply street address for UPS shipping)
____

City: ______________ State/Province: ______________

Zip/Postal Code: ______________ Country: ______________

Tel: (   ) ______________________________

Fax: (   ) ______________________________

Email Address: ______________________________

Qty.	Product Code	Title	Price

❑ **New Edition No Obligation Standing Order Program**
Please enroll me in this program for the products I have ordered. Government Institutes will notify me of new editions by sending me an invoice. I understand that there is no obligation to purchase the product. This invoice is simply my reminder that a new edition has been released.

Subtotal __________

MD Residents add 5% Sales Tax __________

Shipping and Handling (see box below) __________

Total Payment Enclosed __________

15 DAY MONEY-BACK GUARANTEE

If you're not completely satisfied with any product, return it undamaged within 15 days for a full and immediate refund on the price of the product.

Within U.S:	**Outside U.S:**
1-4 products: $6/product	Add $15 for each item (Airmail)
5 or more: $3/product	Add $10 for each item (Surface)

Now Order Online: www.govinst.com

SOURCE CODE: BP01